Guidelines for Process Safety
During the Transient Operating Mode:

Managing Risks during Process Start-ups and Shut-downs

Guidelines for Process Safety
During the Transient Operating Mode:
Managing Risks during Process Start-ups and Shut-downs

Center for Chemical Process Safety

of the

American Institute of Chemical Engineers

New York, NY

Registered Office
John Wiley & Sons, Inc., 111 River Street, Hoboken, NJ 07030, USA

Editorial Office
111 River Street, Hoboken, NJ 07030, USA

For details of our global editorial offices, customer services, and more information about Wiley products visit us at www.wiley.com.

Wiley also publishes its books in a variety of electronic formats and by print-on-demand. Some content that appears in standard print versions of this book may not be available in other formats.

Library of Congress Cataloging-in-Publication Data Applied for:
ISBN: 9781119529156

Cover Design: Wiley-AIChE
Cover Images: Silhouette, oil refinery © manyx31/iStock.com,
Stainless steel © Creativ Studio Heinemann/Getty Images,
Dow Chemical Operations, Stade, Germany/Courtesy of The Dow Chemical Company

SKY10022809_112320

This book is one in a series of process safety guidelines and concept books published by the Center for Chemical Process Safety (CCPS). Refer to www.wiley.com/go/ccps for full list of titles in this series.

It is sincerely hoped that the information presented in this document will lead to an even more impressive safety record for the entire industry; however, neither the American Institute of Chemical Engineers, its consultants, CCPS Technical Steering Committee and Subcommittee members, their employers, their employers' officers and directors, nor ioMosaic Corporation, and its employees and subcontractors warrant or represent, expressly or by implication, the correctness or accuracy of the content of the information presented in this document. As between (1) American Institute of Chemical Engineers, its consultants, CCPS Technical Steering Committee and Subcommittee members, their employers, their employers' officers and directors, and ioMosaic Corporation and its employees and subcontractors, and (2) the user of this document, the user accepts any legal liability or responsibility whatsoever for the consequence of its use or misuse

Table of Contents

Contents

Part I—Normal Operations

Contents   xi

List of Figures

List of Tables

Acronyms and Abbreviations

AIChE	American Institute of Chemical Engineers
C	Consequence (*see Equation 10.1*)
CCPS	Center for Chemical Process Safety
ERP	Emergency Response Plan
ERT	Emergency Response Team
ESD	Emergency Shutdown Device
ESS	Emergency Shutdown System
F	Frequency (*see Equation 10.1*)
HAZOP	Hazards and Operability Study
HIRA	Hazards Identification and Risk Analysis
ITPM	Inspection, Testing, and Preventive Maintenance program
LOPA	Layer of Protection Analysis
LL	Lesson Learned
LS	Lesson Strength
LTI	Lost Time Injury
MOC	Management of Change
MTC	Medical Treatment Case
OD	Operational Discipline (*see Equation 10.1*)
ORR	Operational Readiness Review
PDCA	Plan, Do, Check, Act
PHA	Process Hazards Analysis
PPE	Personal Protective Equipment
PSSR	Pre-Startup Safety Review
R	Risk (*see Equation 10.1*)
RBPS	CCPS Risk Based Process Safety

SIMOPS	Simultaneous Operations
SIS	Safety Instrumented System
SWP	Safe Work Practice
US CSB	United States Chemical Safety and Hazard Investigation Board

Glossary

This Glossary contains some of the common terms in this guideline and some of the relevant process safety-related terms from the CCPS Process Safety Glossary. Refer to Table 2.1 for the glossary terms specific to normal, abnormal, and emergency operations, and to Table 2.2 for those specific to the transient operating modes. The terms in this guideline are current at the time of publication; refer to the CCPS website for updates to the CCPS Process Safety Glossary.

Term	Definition
Conduct of Operations (COO)	An element in the CCPS RBPS model. The embodiment of an organization's values and principles in management systems that are developed, implemented, and maintained to (1) structure operational tasks in a manner consistent with the organization's risk tolerance, (2) ensure that every task is performed deliberately and correctly, and (3) minimize variations in performance. See Operational Discipline.
Consequence (C)	The undesirable result of the impact of a loss event, usually measured in health and safety effects, environmental impacts, loss of property, and business interruption costs. See Equation 10.1.
Deviation	A process condition outside of established design limits, safe operating limits, or standard operating procedures.
Emergency Shutdown Device (ESD)	A device that is designed to shut-down the system to a safe condition on command from the Emergency Shutdown System.

Term	Definition
Emergency Shutdown System (ESS)	The safety system that overrides the action of the basic control system when predetermined conditions are violated.
Frequency (F)	Number of occurrences of an event per unit time (e.g., 1 event in 1000 yr. = $1 \times 10\text{-}3$ events/yr.). Interchangeable with "likelihood". See Equation 10.1.
Hazards Identification and Risk Analysis (HIRA)	A collective term that encompasses all activities involved in identifying hazards and evaluating risk at facilities, throughout their life cycle, to make certain that risks to employees, the public, or the environment are consistently controlled within the organization's risk tolerance.
Inspection, Testing, and Preventive Maintenance (ITPM) Program	Scheduled proactive maintenance activities intended to (1) assess the current condition and/or rate of degradation of equipment, (2) test the operation/functionality of equipment, and/or (3) prevent equipment failure by restoring equipment condition.
Loss event	Point in time in an abnormal situation when an irreversible physical event occurs that has the potential for loss and harm impacts.
Operational Discipline (OD)	The performance of all tasks correctly every time.

Note: OD is the execution of the CCPS RBPS Conduct of Operations (COO) element by |

Term	Definition
	individuals within the organization. See effect on Risk in Equation 10.1.
Outage	See Turnaround.
Pickling	A metal surface treatment used to remove impurities, such as inorganic contaminants, rust or scale from ferrous metals, copper, precious metals and aluminum alloys. A solution called pickle liquor, which usually contains acid, is used to remove the surface impurities.
Plan, Do, Check, Act (PDCA)	A four-step process for quality improvement. In the first step (Plan), a way to bring about improvement is developed. In the second step (Do), the plan is carried out. In the third step (Check), what was predicted is compared to what was observed in the previous step. In the last step (Adjust), plans are revised to eliminate performance gaps.
Risk	A measure of human injury, environmental damage, or economic loss in terms of both the incident likelihood (frequency) and the magnitude of the loss or injury. *Note:* A simplified version of this relationship expresses risk as the product of the frequency and the consequence's impact (i.e., Risk = Frequency x Consequence). See Equation 10.1.

Term	Definition
Risk Based Process Safety (RBPS)	The Center for Chemical Process Safety's (CCPS) process safety management system approach that uses risk-based strategies and implementation tactics that are commensurate with the Risk Based Process Safety (no hyphen) need for process safety activities, availability of resources, and existing process safety culture to design, correct, and improve process safety management activities.
Risk matrix	A tabular approach for presenting risk tolerance criteria, typically involving graduated scales of incident likelihood on the Y-axis and incident consequences on the X-Axis. Each cell in the table (at intersecting values of incident likelihood and incident consequences) represents a particular level of risk.
Safety Instrumented System (SIS)	A separate and independent combination of sensors, logic solvers, final elements, and support systems that are designed and managed to achieve a specified safety integrity level.
	Note: A SIS may implement one or more Safety Instrumented Functions (SIFs).

Term	Definition
Shutdown (S/D)	A process by which an operating plant or system is brought to a safe and non-operating mode.

Note for this guideline: Equivalent to the multiple shut-down transient operating modes discussed in this book (Table 2.2). |
| Significant loss event | Point in time in an abnormal situation when an irreversible physical event occurs that causes loss and harm impacts.

Note: Companies can specify "significant" depending on the relative impact to people, the environment, property, and business interruption—their "risk tolerance." See Risk matrix. |
| Standby mode | Hardware operation that is normally not running but should be ready to run, e.g., an emergency diesel generator.

Note for this guideline: When troubleshooting equipment during an abnormal situation, specific equipment may be placed in a temporary "standby" or "warm circulation" mode while the troubleshooting efforts are underway. |
| Turnaround | A scheduled shutdown period when planned inspection, testing, and preventive maintenance (ITPM), as well as corrective maintenance such as modifications, replacements, or repairs is performed.

Note: Often referred to as a shutdown (Per this guideline: no hyphen) or an outage. |

Acknowledgments

The American Institute of Chemical Engineers (AIChE) and the Center for Chemical Process Safety (CCPS) express their appreciation and gratitude to all members of the *Guidelines for Process Safety During the Transient Operating Mode* Subcommittee for their generous efforts in the development and preparation of this important guideline. CCPS also wishes to thank the subcommittee members' respective companies for supporting their involvement during the different phases in this project.

Subcommittee Members (2016-2018):

Theresa Broussard, Chair	Chevron Corporation
Susan Bayley	Linde Process Plants, Inc.
Don Connolley	BP Corporation NA, Inc.
Eddie Dalton	BASF Corporation
Reyyan Koc Karabocek	ExxonMobil
Pamela Nelson	Solvay
Jitesh Patel	New Jersey DEP
Scott Schiller	BakerRisk
Dan Sliva	CCPS Staff Consultant
Sandeep Vipra	Reliance
Elliot Wolf	Chemours

The collective industrial experience and know-how of the subcommittee members makes this guideline especially valuable to those who develop and manage process safety programs and management systems.

Before publication, all CCPS guidelines are subjected to a peer review process. CCPS gratefully acknowledges the thoughtful comments and suggestions of the peer reviewers. Their work enhanced the accuracy and clarity of this guideline.

Initial manuscript peer reviewers (2018):

Daniel Callahan	Stepan
Charles Foshee	Chevron
Jennifer Mize	Eastman
Al Morrison	Chevron

Although the peer reviewers provided comments and suggestions, they were not asked to endorse this guideline and did not review the final manuscript before its release.

The book committee wishes to express their appreciation to Elena Prats and Kathy Anderson, ioMosaic Corporation, for their contributions in preparing the guideline's draft manuscript. Sincere appreciation is extended to Dr. Bruce K. Vaughen, PE, CCPSC, of CCPS for his contribution in restructuring the book committee's efforts, addressing the final comments from both the book's committee and peer reviewers, and in creating the final, published manuscript.

Before publication of the final manuscript for this guideline, an additional technical review of each of the chapter's drafts and the restructured and enhanced manuscript provided additional insights that were incorporated into the final, published manuscript. Much appreciation is extended to the final reviewers for their time.

Restructured draft and final manuscript reviewers (2020):

Theresa Broussard, Chair	Chevron Corporation
Dan Sliva	CCPS Staff Consultant
Jennifer Bitz	CCPS, Project Manager
Pete Lodal	Eastman
Dr. Anil Gokhale, PE	CCPS Project Director

Special appreciation is extended to Kiezha Ferrell for copy editing the draft of the final manuscript.

Dedication

This guideline is dedicated to those who have been injured or lost their lives, and their families, as a result of incidents that occurred during transition times in the industries handling hazardous materials and energies. Hazards and risk management approaches continue to improve facilities based on the learnings from these incidents. Since there is a better understanding of some of the human performance-related issues today, improved engineering and administrative controls have been established to help prevent these type of incidents from occurring again. This guideline puts forth a framework that applies the Risk Based Process Safety (RBPS) principals to activities specifically associated with the transient operating mode. It is the hope of the authors that the risk reduction approaches provided in this guideline can be integrated and implemented within the existing process safety and risk management systems in organizations striving toward world-class process safety performance.

> *"Do not let the world's adversity stifle your enthusiasm, nor blind your vision. The struggle towards excellence should ever be conducted on the high plains of self-confidence, a sense of purpose, and positive thought. Go forth with a desire to accomplish...with a desire to contribute to our society."*
>
> Ronald E. McNair, 1950 - 1986
> NASA Astronaut - Space Shuttle Challenger

Foreword

As the Chairperson of the U.S. Chemical Safety and Hazard Investigation Board (CSB), which independently investigates the root causes of incidents involving hazardous substance releases, I have seen the complex processes involved in transient operations. More importantly, I, and many dedicated safety professionals, have borne witness to the human cost, property damage, and environmental destruction arising from infrequent, non-routine procedures used at company facilities.

The CSB previously has examined transient operations issues in its investigations, some of which are discussed in this book, including emergency shut-down procedures, incidents resulting from start-up activities, and incidents resulting from normal shut-down activities. For example, lessons learned from start-up operations have shown that facilities should follow established start-up procedures, carefully perform pre-start-up safety reviews and thoroughly check equipment, piping, instrumentation and tanks for damage and safety system impairments. In the case of performing tasks that are not part of normal processing, and often those tasks that are completed under manual control or with less- frequent training, due diligence is necessary. Preparation and sound management systems are essential to ensure safe, predictable transient operations. Good safety practices are good business practices.

Beyond process or facility outages and decommissioning activities, companies face unplanned shut-down and start-up activities from extreme weather, malicious intent, and other external forces. The 2017 hurricane season in the United States was one of the most destructive in the country's history, requiring many companies to stop their operations temporarily, which would require them to start-up again. Likewise, the world continues to be mindful of the risk of having unplanned or emergency outages due to malicious actors' behaviors. These unexpected, complex challenges in managing chemical process hazards will remain a consistent consideration for the foreseeable future.

Yet, safety professionals also recognize that no single checklist is adequate to conduct a thorough risk assessment. Hazard identification is a critical first step to managing safety risks enough to prevent catastrophic, high impact incidents. CCPS and its dedicated expert volunteers and staff continuously strive to collect and share lessons in process safety. Their publications serve as one foundational element of process safety training and implementation programs. These guidelines are the next meaningful contribution to the chemical industry to enable facilities to follow established procedures and checklists related to transient operations.

American philosopher William James said, "To study the abnormal is the best way to understand the normal." This book provides a valuable perspective into both the special characteristics of abnormal and unexpected facility operations. Facilities have the ability to adhere to appropriate safety management systems, engage in emergency planning and response activities in the event of an incident, and to learn from near misses or minor incidents to continually improve.

The challenges in information-sharing and information assimilation within the industry are both difficult and ever-changing. Safety professionals are working on these challenges with enthusiasm, tenacity, and dedication to develop better methods of analysis and implementation of robust safety practices. The CCPS and its Technical Steering Subcommittee have done a tremendous job reflecting that tenacity by providing new perspectives about incident prevention, hazard identification, and mitigation efforts to address the vulnerability during transient operations. In this new age of technical and operational interconnectivity and interdependence, it is necessary to provide broad lessons to practitioners, both working professionals and students, with comprehensive guidance on planned and unplanned start-up or shut-down activities. The rarity with which these activities may occur does not reduce the catastrophic human and environmental impact of an inadvertent chemical release, fire, or explosion.

Continual improvement is a core value at the CSB, and it is defined as creating a culture that seeks to learn from all experiences, acquires new knowledge, considers all viewpoints, sets new goals, and applies lessons learned. If it is true, as Mark Twain once said, that "continuous improvement is better than delayed perfection", then readers will enjoy the opportunity to delve into these guidelines to do their work right the first time. Notably, the guidelines and its appendix make it much easier to assimilate detailed (albeit non-exhaustive) lists of many potential hazards and risks inherent during transient operations.

Safety is a shared responsibility, and safe transient operations require planning, communication, good procedures, training, interdisciplinary worker involvement and great implementation. This book provides a framework for any facility processing hazardous materials to reduce the number of incidents, of any severity, during transient operations.

Vanessa Allen Sutherland
Chairperson and Chief Executive Officer, 2015-2018
U.S. Chemical Safety and Hazard Investigation Board
Washington, D.C. USA

Preface

The American Institute of Chemical Engineers (AIChE) has been closely involved with process safety, environmental, and loss control issues in the chemical, petrochemical, and allied industries for more than four decades. Through its strong ties with process designers, operators, safety professionals, and members of academia, AIChE has enhanced communications and fostered continuous improvement between these groups. AIChE publications and symposia have become information resources for those devoted to process safety, environmental protection, and loss prevention.

AIChE created the Center for Chemical Process Safety (CCPS) in 1985 soon after the major Industrial incident in Bhopal, India, in 1984. The CCPS is chartered to develop and disseminate technical information for use in the prevention of incidents. The CCPS is supported by more than 210 industry sponsors, who provide the necessary funding and professional guidance to its technical steering committees. The major product of CCPS activities has been a series of guidelines and concept books to assist those implementing various elements of the Risk Based Process Safety (RBPS) approach. This guideline is part of that series.

The scope of this guideline addresses process safety activities that are essential for effectively managing the risks associated the different transient operating modes, recognizing that not all activities will apply to every mode. The intended audience for this guideline is everyone in an organization who should ensure process safety: managers, supervisors, engineers, operators, mechanics, electricians, and associated support personnel. This guideline reinforces why everyone needs to have the operational discipline to understand their role and apply their skills to complete their tasks competently; it helps reduce the process safety risks, especially during the activities associated before, during, and after a process transitions between the idle, at-rest, and safe state and its normal operation.

The guideline defines the transient operating mode in context of normal, abnormal, and emergency operations to provide a clear and understandable terminology and to establish a consistent language for conveying the risks associated with these process start-ups and shut-downs and how these risks can be effectively managed during these times. In addition, this guideline will provide brief overviews of methodologies for activities associated with the transient operating mode that are covered in more detail in other CCPS publications. One example is the connection between this guideline and another CCPS guideline when integrating process safety into engineering projects associated with process and facility shutdowns (the details are provided in Chapter 4 and Chapter 5).

Process safety should be a major consideration for establishing safe processing condition-related transitions in the chemical, petrochemical, and associated industries. Understanding how best to manage the risks associated with the transient operating mode, especially when starting up or shutting down a process, will help reduce the risks and help prevent and mitigate incidents. The CCPS Technical Steering Committee initiated the creation of this guideline to assist companies striving toward world-class process safety performance, especially during start-ups and shut-downs.

> *"Safety, loss prevention, security, and environmental protection are the basic responsibilities of the chemical engineer both as an engineer and as a supervisor or manager of chemical operations and marketing. The engineering student and professional should make them primary considerations in both study and practice of the profession."*
> Kent Davis
> Former Director of Personal and Safety
> Dow Chemical U.S.A. Operations (ca. 1988 [1])

1 Introduction

1.1 Introduction

This chapter discusses the scope of, the audience for, and the benefits for the readers of this guideline. For readers unfamiliar with the CCPS Risk Based Process Safety (RBPS) approach, this chapter also includes a brief overview of its framework, including how lessons learned from experience (one of the RBPS pillars) are incorporated into each chapter. The last section in this chapter provides the reader with the guideline's framework: how the chapters are organized based on the types of operations at a facility (normal, abnormal, and emergency) and how the risks associated with each transient operating mode—the subject of this book—depends on which mode of operation the process is undergoing at that time.

1.2 Scope

The scope of this guideline addresses process safety activities that are essential for effectively managing the risks associated with the different *transient operating modes*, recognizing that not all activities will apply to every mode. Since the risk of incidents can be high during the start-ups and shut-downs for normal operations in most manufacturing facilities, this book presents incidents that occurred during start-ups and shut-downs, providing insights as to why they happened and guidance on how to minimize the risk in the future. The important distinction between "transient operations" and the "transient operating mode" should be understood. This guideline defines the transient operating mode in the context of normal, abnormal, and emergency operations, providing a clear and understandable terminology of what is a "transient operating mode." Details on this distinction will be provided in Chapter 2.

The other "transient operations" *do not apply* to the scope of this guideline. These transient operations include, among others, "infrequent or non-routine activities" and "workarounds." The distinction is critical for understanding of this guideline's scope, as the term "transient operations" means many different things depending on the context and views of those who are noting such transition times.

For example, "transient operations" include a major grade change in a polymer reactor or summer to winter operating mode change for a refinery. These tend to be routinely practiced, have well documented procedures and are usually done without a shut-down. These types of transient operations are not covered in this guideline either.

It should be noted that the hazards and risks associated with non-routine activities, performed less frequently than normal activities, should be understood by everyone performing the infrequent task. As discussed in this guideline's Foreword, "many dedicated safety professionals...have borne witness to the human cost, property damage and environmental destruction..." when non-routine, infrequent tasks are not performed safely.

In addition, workarounds can be defined as "creative solutions made to overcome an issue without actually solving it." Well-intentioned workarounds can be small, quick, and simple "one-minute" changes which create "bad memories" at a facility [2, pp. 209-234]. In other words, incidents have occurred – and will occur - when those performing the infrequent tasks or making simple material or procedural changes have not thoroughly refreshed their understanding of the task's hazards and have not thoroughly addressed the task's risks.

Another way to describe the scope of this guideline is to recognize that transient operating modes are planned, anticipated activities, even though some of them, such as an emergency shut-down, are prevented as much as possible. Whether during normal operations or when in transition, all operating modes need to have developed procedures which are documented, reviewed, and practiced through training and on-the-job exercises. By contrast, there are other transient operating modes that are not planned, and therefore should be handled ad hoc by use of rigorous change management and hazards assessment procedures for the specific situation at hand. For example, a batch reactor that has been fed the wrong reactant by mistake and has created a viscous material which cannot be removed by normal cleaning procedures would not have a written procedure applicable to this situation. In this case, a team of experts would be needed to develop specific, non-routine actions to recover from this deviation. This team of experts would understand potential hazards and assess their associated risks, ensuring that the non-routine activity will not place people at risk to a loss of control of a hazardous material or energy.

In short, if the abnormal operation has a procedure associated with it (even if it is only "hit the shut-down button and evacuate"), it is covered by this guideline. If not, then the activity is outside the scope.

Clear definitions help establish a consistent language between practitioners when discussing the transient operating mode-specific risks and how these risks can be effectively managed for each of the transient operating modes. Since the scope of this guideline is limited to the transient operating mode, other references are noted as additional resources [3] [4] [5] [6] [7] [8] [9]. In particular, some of the incident data summarized from the 1990's includes the following:

> *"...emphasizes the need for start-up and shut-down procedures to be in sufficient detail to cover known pitfalls and contingencies. In*

the case of maintenance, good communication is essential so than everyone knows what others are doing at any time and what responsibilities each has. Split or unclear responsibilities are a recipe for disaster. [3, pp. 3-4] "

"To [help] prevent these types of incidents from occurring, facilities should employ effective communications, provide workers with appropriate training, and have in place strong and up-to-date policies and procedures for hazardous operations such as start-ups and shut-downs [5, p. 1]."

When the equipment is being shut down (the transition time), the word *shut-down* is used in this guideline. This spelling, with the hyphen between the words, distinguishes the activities taken on operating equipment *during a transition* (when the equipment is being *shut-down*) and the activities conducted when the equipment *is not operating* (during a *shutdown* – no hyphen). In particular, the hyphen is essential when discussing incidents, as many incident reports designate "shutdown" without distinguishing between the shut-down mode and a project-related shutdown. Thus, the incidents in this guideline will focus on those that occurred when the equipment *was being shut-down*. For consistency within this guideline as well, when the process equipment or process is starting up, *start-up* with a hyphen will be used (recognizing that another common spelling is "startup").

This guideline will discuss two general transient operating modes (the time when a process is in transition between its idle and its operating states.) Specifically, the normal start-up and normal shut-down times are defined as:

1. Start-up time—from an idle, safe, and at-rest state to normal operations, and
2. Shut-down time—returning from normal operations to its normal idle, safe, and at-rest state.

If there is an emergency shut-down, the controlled end state of the process equipment should be idle, safe, and at-rest. If the emergency shut-down is unanticipated, the non-operating end state may not be safe. Any hazards and their associated risks should be addressed before resuming operations.

Every effort has been made to be consistent throughout this guideline when distinguishing between the start-up and shut-down modes and the startup and shutdown stages in a project (see discussion in Chapter 4).

1.3 Audience

The intended audience for this guideline are those at all levels in an organization who should ensure process safety, especially during the transient operating mode. This includes decision-makers at all levels, including: senior executives, corporate managers, facility and group managers, supervisors, engineers, operators, mechanics, electricians, and associated support personnel, as well as contractors and those managing contractors. *This audience is not too broad* [10] [11] [12] [13]. Severe incidents have occurred when decision-makers, no matter what their level in an organization, were not aware of how their decision could—and did—increase their process safety risks during the transient operating mode.

Everyone should understand how their decisions can adversely affect the risks associated with a process undergoing one of its two transition times: its start-up or its shut-down. When everything is going well, when the normal start-ups, operations, and shut-downs proceed as expected, the risks can be managed effectively based on operating experience, well-established operating procedures, and thorough skills-based training. These efforts across the entire organization represent part of a comprehensive process safety and risk management program. Thus, everyone has the operational discipline to follow the established procedures and understand how to safely manage unexpected situations which may arise during transition times. The following pages show that without understanding of the complexity inherent in managing the process safety risks, some quick decisions have led to serious process safety incidents. These incidents occurred, in part, due to both short term and long term decisions in different groups, such as those resourcing the manufacturing efforts, those scheduling the product, and those designing, constructing, operating, maintaining, and changing the equipment handling hazardous materials and energies.

1.4 Benefits

It is hoped that the information presented here, gleaned from decades of process safety hazard and risk reduction experiences around the world, and experiences both good and abysmal, has been adequately compiled and conveyed on these pages. In particular, when this information is understood and applied to address the risks associated with the start-up and shut-down times, the overall risk to an organization will be reduced. The significant consequences and incident's impact will be reduced; there will be fewer incidents with fatalities, serious injuries, environmental harm, property damage, and business interruption losses. Those are the benefits.

1.5 Applying CCPS Risk Based Process Safety (RBPS)

The CCPS Risk Based Process Safety (RBPS) approach is a proven methodology to help organizations understand their process safety hazards and effectively manage their process safety risks [14]. The RBPS approach consists of a process safety management program with four "pillars" and twenty "elements" that integrate a company's overall process safety risk management efforts. This approach was compiled from successful experience in many industrial organizations across a broad range of industries in different jurisdictions around the world. The approach continues to evolve today. For readers unfamiliar with these pillars and elements, refer to Chapter 10 for an overview, as the discussions and lessons learned from start-up and shut-down incidents will be based on the CCPS RBPS foundation. In addition, note that there is no hyphen—*by design*—when the "CCPS Risk Based Process Safety (RBPS)" approach is being discussed. However, a Risk-Based Inspection (RBI) program, for example, used in maintenance-related efforts applies a hyphen between risk and based.

1.6 Incident discussions and guidance

This guideline uses incidents, from both published investigation reports and internal company incident information, that provide details on what went well and what went wrong during the start-up or shut-down. The anonymous company incidents submitted to this book or located in generic incident databases are presented for sharing.

Everyone learns from experience.

The goal of sharing incidents is to prevent others from learning from the bad experience the hard way. The collective global goal is to reduce the process safety risks and prevent incidents that cause harm to people, the environment, and the business. As the cases presented are reviewed, it should be noted that:

1. The guidance—these learnings—are framed within the CCPS Risk Based Process Safety (RBPS) approach described in Chapter 10,

And most importantly:

2. The year of these cases is noted since those that occurred *before* the publication of the initial CCPS RBPS guidance in 2007

were typically used, in part, to help *provide the learnings from the industrial experience reflected in this approach to risk.*

Thus, these incidents provide learnings by identifying the elements that were not implemented or were ineffectively implemented at the time of the incident. Unfortunately, the incidents which have occurred after 2007 only reinforce the simple fact that there is still much to learn. Even with later publications of other risk-based approaches, the significant incidents might not have happened had the company understood and applied a risk-based process safety approach.

1.7 Framework

The overall chapter order in this guideline follows the operations team activities using the framework based on the three general operating modes—normal, abnormal, and emergency—depicted in Figure 1.1:

- Introduction and definitions (Chapters 1 and 2)
- Part I – Normal Operations Chapters 3, 4, and 5)
- Part II – Abnormal and Emergency Operations (Chapters 6, 7, and 8)
- Part III – Other Considerations (Chapters 9 and 10)
- Appendix (Incident Reviews and Additional Information)

These three operating modes set the stage for different types of transition times. It is essential to note that normal operations are the day-to-day, expected mode of operation. During this mode, everything goes as expected. There are standard start-up and shut-down procedures for transitioning between idled and operating equipment—everything goes according to design and plan. However, not everything goes according to plan every day. Thus, the operations team may have abnormal operations, or if matters cannot be resolved, may have emergency operations.

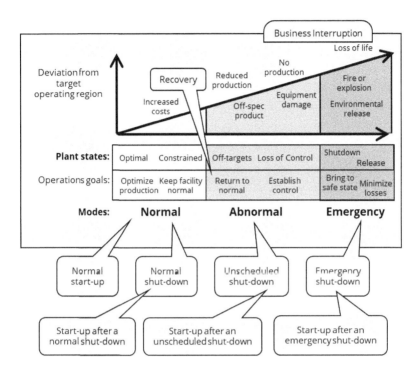

Figure 1.1 Three types of facility operations and their corresponding
transient operating modes.
(Adapted from [15, p. 22])

The ten transient operating modes discussed in this guideline are introduced in Table 1.1. The Appendix provides the summary of a detailed incident review focusing on published incidents that occurred during the transient operating modes listed in Table 1.1, and includes additional guidance on how to more effectively manage unexpected situations, especially those than may cause or may happen during transient operating modes.

Table 1.1 List of the different types of transient operating modes.

Different Types of Transient Operating Modes	
1	Shut-down (normal)
2	Start-up (normal)
3	Shut-down designed for a planned shutdown
4	Start-up after a planned shutdown
5	Shut-down designed for an extended shutdown
6	Start-up after an extended shutdown
7	Shut-down activated for an unscheduled shutdown
8	Start-up after an unscheduled shutdown
9	Shut-down activated for an emergency shutdown
10	Start-up after an emergency shutdown

The corresponding chapters—the guideline's framework—are listed in Table 1.2:

1. The scope of this guideline, the ten start-up and shut-down transient operating modes, are covered in Chapter 4 through Chapter 8, and
2. Other transition times specifically associated with the equipment or process life cycle, the *initial* start-ups and the *final* shut-downs, are covered in Chapter 9.

There are times when the equipment is scheduled for special projects or routine maintenance [16] [17]. Although these project- or maintenance-related activities are performed safely on idled equipment, sometimes these activities are performed at or near operating equipment. The normal operations flow chart (Figure 1.2), including the transition times—the transient operation modes—and associated process and facility shutdowns, covers the normal operations mode; Chapter 3 discusses the normal operations; Chapter 4, process shutdowns; and Chapter 5, facility shutdowns. The transient operating modes—the shut-downs and start-ups—associated with normal operations, process shutdowns, and facility shutdowns are covered in each of these chapters, as well. These modes of operation will be defined in more detail in Chapter 2.

Table 1.2 Chapter framework for this guideline.

Chapter		Type of Transient Operating Mode	
1	Introduction		
2	Defining the Transient Operating Mode		
Part I - Normal Operations			
3	Normal Operations	1	Shut-down (normal)
		2	Start-up (normal)
4	Planned Shutdowns	3	Shut-down designed for a planned shutdown
		4	Start-up after a planned shutdown
5	Extended Shutdowns	5	Shut-down designed for an extended shutdown
		6	Start-up after an extended shutdown
Part II - Abnormal and Emergency Operations			
6	Recovery Operations		
7	Unscheduled Shutdowns	7	Shut-down activated for an unscheduled shutdown
		8	Start-up after an unscheduled shutdown
8	Emergency Shutdowns	9	Shut-down activated for an emergency shutdown
		10	Start-up after an emergency shutdown
Part III - Other Considerations			
9	Other Transition Time Considerations		
10	Risk Based Process Safety (RBPS) Considerations		
Appendix			
A	Incident Review		

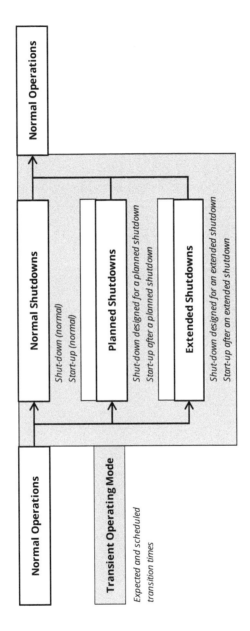

Figure 1.2 Example normal operations flow chart.

However, process upsets are anticipated and accounted for with engineering and administrative controls designed and implemented to monitor and respond to these upsets. These abnormal operations can have successful or unsuccessful outcomes, either recovering the process back to its normal operating conditions and resuming normal operations or resulting in an unscheduled process shut-down. As is depicted in Figure 1.3, the equipment and process may need to shut-down when the recovery efforts are unsuccessful, transitioning the process from its normal operating state to the shutdown mode. It is important to note at this point that there is a temporary operating time, often designated as a standby mode, which provides time for process upsets to be diagnosed before either completely recovering the process conditions or shutting the process down.

As is depicted in Figure 1.3, there are two general types of shut-downs during abnormal operations: normal shut-down procedures and emergency shut-down procedures. Note that normal shut-down procedures can be used for both planned and unplanned shut-downs. Emergency shut-down procedures are, by definition, unplanned events and emergency situations. If a significant loss event occurs, the emergency shut-down procedures are activated, potentially resulting in impact to people, the environment, and assets.

Part II of this guideline covers the abnormal and emergency operations modes: Chapter 6 discusses the successful response to abnormal operations—the recovery; Chapter 7, the shut-downs and start-ups afterward associated with unscheduled shutdowns; and Chapter 8, the shut-downs and start-ups afterward associated with emergency shutdowns.

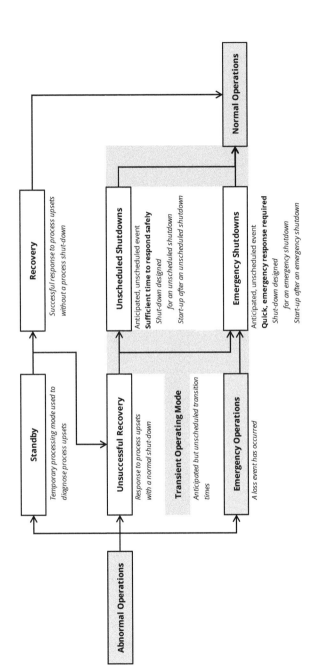

Figure 1.3 Example abnormal and emergency operations flow chart.

Part III of this guideline covers other items that should be considered when effectively managing the risks associated with transition times. Chapter 9 covers the transition times related to the facility life cycle, such initial start-ups and decommissioning—the steps essential for a final shut-down. The last chapter, Chapter 10, provides a brief overview of the CCPS Risk Based Process Safety (RBPS) approach, providing guidance on how the RBPS elements apply during transition times.

2 Defining the Transition Times

2.1 Introduction

This brief chapter summarizes the definitions of the three modes of operation (Section 2.2) and the ten transient operating modes (Section 2.3) as they apply to this guideline. These definitions apply to the specific transient operating modes discussed in the following chapters. This chapter concludes with a start-up incident, framing the issue first—which transient operating mode applies—to help understand what went wrong and then make systemic changes to help prevent it from happening again.

2.2 Defining the modes of operation

As was shown in Figure 1.1, there are three types of facility operations: normal, abnormal, and emergency. The definitions for *modes of operation* used in this guideline are listed in this order in Table 2.1, based on when they are introduced and then discussed in more detail in each chapter:

- Modes of Operations:
  - Normal Operations (*Mode*)
  - Abnormal Operations (*Mode*)
  - Emergency Operations (*Mode*)
  - Transient Operations *(Mode)*
- Transient Operating Mode
- Shutdown Mode
- Process Shutdown Mode
- Facility Shutdown Mode
- Abnormal Situation or Mode
- Recovery Mode

2.3 Responses to deviations during operations

The definitions for the ten *transient operating modes* introduced in Table 1.1 are provided in Table 2.2, listed in order on when they are introduced and then discussed in detail in this guideline's chapters:

Part I Normal Operations

1. Shut-down (normal)
2. Start-up (normal)
3. Shut-down for a process shutdown
4. Start-up after a process shutdown
5. Shut-down for a facility shutdown
6. Start-up after a facility shutdown

Part II Abnormal and Emergency Operations

7. Shut-down for an unscheduled shutdown
8. Start-up after an unscheduled shutdown
9. Shut-down for an emergency shutdown
10. Start-up after an emergency shutdown

Each of these transient operating modes will be discussed in Part I and Part II.

Table 2.1 Definitions for the modes of operation.

Normal Operations (Mode)	The operating mode when the process is operating between its start-up and shut-down phases (its transient operating modes) and within its normal operating conditions (its standard conditions).
	For a Continuous Process (*an open, steady-state system*) The time when the process conditions, such as the flow rates, temperatures, and pressures, are *not changing over time* - when the process is at the normal operating conditions and within its standard operating limits.
	For a Batch Process (*a closed, unsteady-state system*) The time when the process conditions, such as temperatures, pressures, and concentrations, *are changing over time* - when the process is at the normal operating conditions and within its standard operating limits.
Abnormal Operations (Mode)	The operating mode that occurs during normal operations when there is a process upset and the process conditions deviate from the normal operating conditions.
Emergency Operations (Mode)	The operating mode that occurs after abnormal operations, when either:
	1) a shut-down is activated due to a significant process upset that exceeds the safe operating limits during abnormal operations.
	2) an emergency shut-down occurs due to a loss event with the loss of containment of hazardous materials or energies during normal operations.
Simultaneous Operations (SIMOPS)	The time associated with simultaneous or adjacent activities, such as drilling and operations or construction and operations, not normally carried out together.
	Note 1: This may introduce novel hazards not anticipated in prior HIRA studies for the facility and bring together staff not familiar with all the risks and how these are managed for the other activity. A formal SIMOPS plan and staff training is usually required to address these aspects safely.
	Note 2: SIMOPS includes work associated with an engineering project or maintenance activity on equipment or processes that are not operating yet (i.e., are in their construction or commissioning stage) or have been taken out of operations for a scheduled inspection, test, or preventive maintenance (ITPM) activity.

Table 2.1 Definitions for the modes of operation (Continued).

Transient Operations (Mode)	The operating mode when the process is in transition and is not in its normal operations mode.
	Note: Transient operations include process start-ups and shut-downs, product transitioning *in between* normal operations, non-routine activities performed *during* normal operations, and activities associated with the project life cycle (e.g., commissioning and start-up of new capital projects, mothballing, and decommissioning).
Transient operating mode	The time when a process is in transition from one operating state to another: when the process is in its *Start-up* or *Shut-down* transient operating mode.
Planned Shutdown (Mode)	The time when the process equipment is not operating and is undergoing a planned project shutdown or a routine maintenance shutdown.
Extended Shutdown (Mode)	The time when the process equipment for entire process units or a facility is not operating and is undergoing a major, planned project-related shutdown.
Abnormal Situation	A disturbance or series of disturbances in a process that cause deviations from the normal processing conditions.
Recovery (Mode)	The time when operations can safely respond to a process upset and return it to normal operations, recovering the process to its normal operating conditions.
Unscheduled Shutdown (Mode)	The time between an unscheduled shut-down and the start-up afterwards.
Emergency Shutdown (Mode)	The time after an emergency shut-down has occurred when the equipment or processes are being inspected, repaired or replaced as needed, all start-up readiness reviews have been completed and the equipment or process is ready for the start-up transient operating mode.

Table 2.2 Definitions for the transient operating modes.

1	Shut-down *or* Normal shut-down	*A transient operating mode:* the time when the process equipment is being shut-down, stopping the equipment and taking the equipment from its normal operating conditions to end normal operations.
		For a Continuous Process *(open, steady-state systems):* A planned series of steps to stop the process at the end of normal operations, taking the process equipment from its normal operating conditions to an idle, safe, and at-rest state.
		For a Batch Process *(closed, unsteady-state systems):* A planned series of steps to stop the process at the end of normal operations, taking the process equipment from its final operating conditions to a clean, idle, safe, and at-rest state in preparation for the next batch.
2	Start-up *or* Normal start-up	*A transient operating mode:* The time when the process equipment is being restarted (or started after being commissioned), taking the equipment to the normal operating conditions for normal operations.
		For a Continuous Process (open, steady-state systems): A planned series of steps to take the process equipment from an idle, safe, and at-rest state to the normal operating conditions.
		For a Batch Process (closed, unsteady-state systems): A planned series of steps to prepare for and begin a batch process, taking the process equipment from a clean, idle, safe, and at-rest state to the normal operating conditions.
3	Shut-down designed for a planned shutdown	*A transient operating mode:* The time that may require procedures in addition to the normal shut-down procedures for stopping the equipment in preparation for a planned project or maintenance shutdown.
		Note: If other groups are involved in the planned shutdown, such as engineering, maintenance, or contractors, special permits and handover procedures must be in place beforehand, as needed, before performing the shutdown-related activities.

Table 2.2 Definitions for the transient operating modes (Continued).

4	Start-up after a planned shutdown	*A transient operating mode:* The time that may require procedures in addition to the normal start-up procedures before restarting the equipment after a planned project or maintenance shutdown.
		Note: If other groups were involved in the planned shutdown, such as engineering, maintenance, or contractors, the special permits and handover procedures implemented for the shutdown-related activities must be reversed, reviewed, and authorized before resuming operations. This includes performing, as needed, equipment Pre-start-up Safety Reviews (PSSR) and Operational Readiness Reviews (ORR).
5	Shut-down designed for an extended shutdown	*A transient operating mode:* The time that requires procedures in addition to the normal shut-down procedures for stopping the equipment in preparation for a major project, a major process unit shutdown, or a major facility turnaround or outage.
		Note: If other groups are involved in the extended shutdown, such as engineering, maintenance, or contractors, special permits and handover procedures must be in place beforehand, as needed, before performing the shutdown-related activities.
6	Start-up after an extended shutdown	*A transient operating mode:* The time that requires procedures in addition to the normal start-up procedures before restarting the equipment after a major project, a major process unit shutdown, or a major facility turnaround or outage.
		Note: If other groups were involved in the extended shutdown, such as engineering, maintenance, or contractors, the special permits and handover procedures implemented for the shutdown-related activities must be reversed, reviewed ,and authorized before resuming operations. This includes performing, as needed, equipment Pre-start-up Safety Reviews (PSSR) and Operational Readiness Reviews (ORR).
7	Shut-down activated for an unscheduled shutdown	*A transient operating mode:* The time when the operations team can use normal or specially-designed shut-down controls and procedures. These shut-downs can occur when:
		1) the process cannot be successfully recovered from an abnormal situation and the normal shut-down procedures can be used
		2) there is time to prepare the facility for a pending natural hazard (e.g., a hurricane or cyclone)
		Note: If other groups are involved in the unscheduled shutdown, such as maintenance, special permits and handover procedures must be in place beforehand, as needed, before performing the shutdown-related activities

Table 2.2 Definitions for the transient operating modes (Continued).

8	Start-up after an unscheduled shutdown	*A transient operating mode:* The time when preparing for and resuming operations after an unscheduled shutdown.
		Note: If other groups were involved in the unscheduled shutdown, such as maintenance, the special permits and handover procedures implemented for the shut-down activities must be reversed, reviewed, and authorized before resuming operations. This includes performing, as needed, equipment Pre-start-up Safety Reviews (PSSR) and Operational Readiness Reviews (ORR).
9	Shut-down activated for an emergency shutdown	*A transient operating mode:* The time when the operations team has to abruptly shut the process down using emergency engineering controls and/or administrative shut-down procedures. These shut-downs can occur when:
		1) the process cannot be successfully recovered from an abnormal situation, when the operating conditions are at, or may have exceeded, the safe operating limits and there has been no loss event
		2) there has been a loss event, requiring that everyone responds safely, including the emergency responders
		3) there has been an unpredictable natural hazard event (e.g., an earthquake, lightning strike).
		Note: Depending on the facility's emergency response resources and on the extent of the loss event, the Emergency Response Team (ERT) and the Emergency Response Plan (ERP) may have to be activated.
10	Start-up after an emergency shutdown	*A transient operating mode:* The time when preparing for and resuming operations after the emergency shutdown period.
		Note: If the emergency shutdown period occurred due to an incident with harm to people, the environment, or property, then all special restart and operations-related projects should be designed and implemented before resuming operations. This includes performing, as needed, equipment Pre-start-up Safety Reviews (PSSR) and Operational Readiness Reviews (ORR).

2.4 A start-up incident

A start-up incident, shown in the illustration in Figure 2.1, resulted in a significant release of an ammonia cloud that drifted across a river [18]. This incident occurred when a roof-mounted pipe failed upon restart of the facility's refrigeration system after a seven-hour power outage. More than 150 people reported exposure to the released ammonia, with thirty-two people being admitted to the hospital and four being placed in intensive care. What happened? Which transient operating mode applies? How can systems be implemented to prevent incidents like this from happening again? The following chapters will explore this incident—and many more—to help understand "what went wrong" and how to prevent incidents from occurring during the transient operating mode. As shown in Figure 2.2, this incident occurred during the start-up after an unscheduled shutdown. This figure provides an overall timeline for helping identify the time of the incident relative to normal, abnormal, and emergency operations and will be discussed in more detail throughout the rest of this guideline. It will be used to determine which transient operating mode applied at the time of the incident. Additional details on this ammonia release incident are described in Chapter 7, Case 7.6.3-1.

Figure 2.1 The anhydrous ammonia release incident at start-up.
[18]

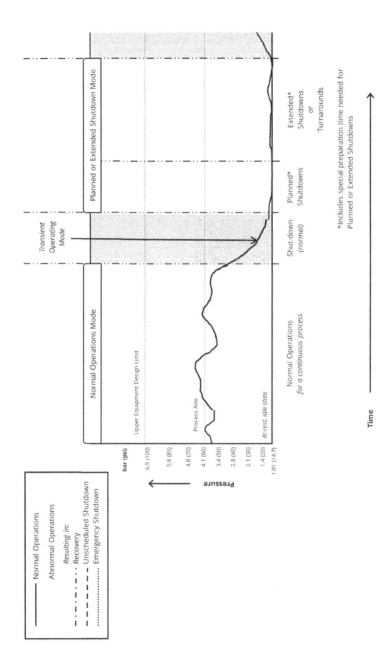

Figure 2.2 Timeline used to determine which transient operating mode applies

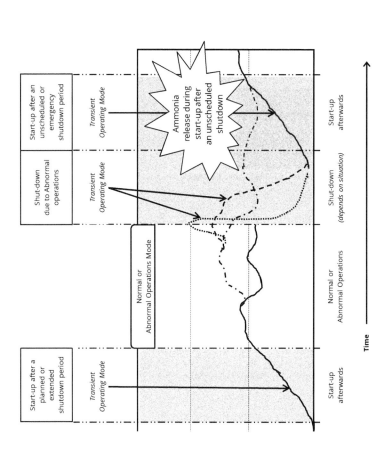

Figure 2.2 Timeline used to determine which transient operating mode applies (continued)

Part I
Normal Operations

3 Normal Operations

3.1 Introduction

This chapter covers the transient operating modes associated with the normal operations associated with the everyday production of materials. Although there are engineering controls designed for day-to-day operation (discussed in Section 3.2), administrative controls are needed, as well, to effectively manage safe start-ups and shut-downs. Since procedures are essential for being able to operate a process consistently and safely, especially during the transient operating mode, this chapter includes a brief overview of procedures in Section 3.3.

This chapter continues with discussions on how normal shut-downs and start-ups are performed safely (Section 3.4 and Section 3.5, respectively). Then some lessons learned from incidents which occurred during normal shut-downs and start-ups are covered in Section 3.6. This chapter concludes with a discussion on the applicable RBPS elements for the normal shut-downs and start-ups in Section 3.7. The next two chapters on process and facility shutdowns tie in to this chapter to complete Part I of this guideline on normal operations.

3.2 The normal operation

Process safety applies to all modes of operations, including normal operations. For this guideline, normal operations are defined in Table 2.1 as:

The operating mode when the process is operating between its start-up and shut-down phases and within its normal operating conditions.

Normal day-to-day operations in the chemical process industries can be continuous, batch, or a combination of the two. Process safety efforts are essential for every day-to-day operation, no matter what type of process it is. For a continuous process, an open, steady-state system, the process conditions during the normal operations time, such as the flow rates, temperatures, and pressures, are not changing over time; the process is at the normal operating conditions and within its standard operating limits. On the other hand, in a batch process, a closed, unsteady-state system, the normal operations time is when the process conditions, such as temperatures, pressures, and concentrations, are *changing* over time; the process is at the normal, but transient, operating conditions and within its standard operating limits. These operating limits are bounded by safety limits, such that deviations from these limits (abnormal operations) must be addressed before a loss event occurs.

Whether continuous or batch, these processes have a normal start-up and a normal shut-down associated with them: either the process is running or not. These processes have normal start-up and shut-down procedures associated with them, as well [19]. This guideline will focus on continuous or batch operations, recognizing that a combination of these guides will apply to processes with have elements of both continuous and batch processes (e.g., a Continuous Stirred Tank Reactor (CSTR)).

This chapter specifically addresses the normal operations mode with normal shut-downs using the procedures dedicated for normal shut-downs and normal start-ups afterwards—transient operating modes Type 1 and Type 2 listed in Table 1.1. Thus the scope of the transient operating modes in this chapter are shown in Figure 3.1. Chapters 4 and 5 cover the special, additional shutdown-related preparations for projects or maintenance-related procedures and their corresponding shut-down and start-up transient operating modes.

A normal operation for a continuous process, focusing on only the pressure of the process, could be represented by a typical curve shown in Figure 3.2; a typical normal operation curve for an isothermal batch process, focusing on the temperature, is depicted in Figure 3.3. There are two transition times for the normal operations: the normal start-up and the normal shut-down, as defined for both types of processes as (Table 2.2):

Normal start-up: The time when the process equipment is being restarted, taking the equipment to the normal operating conditions for normal operations.

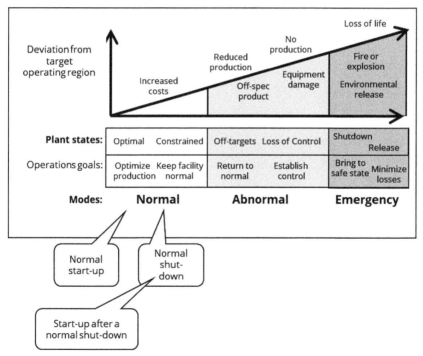

Figure 3.1 Transient operating modes during normal operations. (Adapted from [15])

Normal shut-down: The time when the process equipment is being shut-down, stopping the equipment and taking the equipment from its normal operating conditions to the end of normal operations.

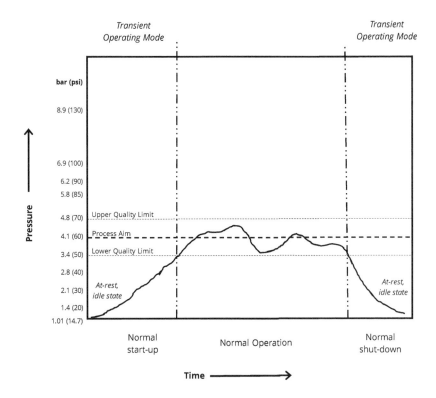

Figure 3.2 Example timeline for a normal operation—continuous process.

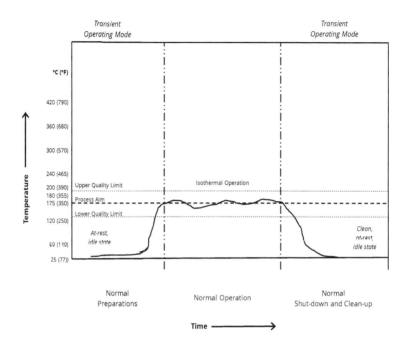

Figure 3.3 Example timeline for a normal operation—batch process.

In particular, for continuous processes, the normal start-up involves the steps taken to move the process equipment from their idle, safe, and at-rest state to the normal operating conditions. For batch processes, the normal start-up is the steps taken to prepare for and begin a batch process, taking the process equipment from a clean, idle, safe, and at-rest state to the normal operating conditions. The normal shut-downs essentially reverse these steps to take the equipment back to the normal safe and at-rest, idle state. Note that equipment failures "during routine operations are sometimes the result of fatigue and stress accumulated in the equipment by ups and downs [8, p. 6]." Hence, it's important to continually monitor the equipment's fitness for use to ensure that the stresses to the equipment over time do not jeopardize the equipment's integrity. In addition, incidents due to unmonitored or unknown corrosion, including rust and corrosion under insulation, are noted in other publications [2] [20]. The corrosion-induced, degraded equipment conditions can contribute to equipment failure during the stresses incurred during a start-up or shut-down.

When they have not been properly maintained, protective equipment failures can occur when called upon in an emergency situation. Although more discussion on effective equipment integrity programs is beyond the scope of this guideline, Inspection, Testing and Preventive Maintenance (ITPM) programs are covered in more detail in other publications [21, pp. 239-260] [22] [23]. A comprehensive listing of protective equipment that should be incorporated in an effective ITPM program is provided elsewhere [20, pp. 250-253].

At this point, it is worth noting that there are engineering and administrative controls which have been designed and are used to monitor and safely respond to process deviations. These process deviations are expected. For example, temperatures can slowly rise or fall, pressures can slowly rise or fall, and flow rates can increase or decrease. It is essential that the process chemistries, technologies, and hazards be understood, and that their risks are assessed and managed. Although the process safety information required for each process will be different, the common responses to process deviations using typical types of engineering and administrative controls, such as alarms, interlocks, and operator intervention, are much the same. These responses, relative to the process intent, are illustrated in Figure 3.4. The ideal normal operation assumes that there are no significant deviations from the process intent; the process will simply proceed from left to right: a normal start-up, a normal production run, then a normal shut-down (continuous, Figure 3.2 and batch, Figure 3.3).

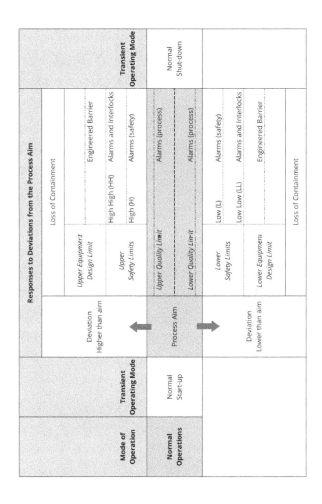

Figure 3.4 Types of engineering and administrative control responses to process deviations.

During abnormal operations, expected process deviations above or below the aim may approach or exceed the upper or lower *safety limits*. However, if process control is not recovered, these deviations may approach or exceed the upper or lower *design limit* of the equipment. When the operating conditions exceed these equipment design limits, catastrophic equipment failure may occur, releasing the equipment's contents (a "loss of containment" event). Such loss events on equipment handling hazardous materials and energies could lead to fatalities, injuries, environmental harm, and property damage. These responses to abnormal operations, introduced with the flow chart illustrated in Figure 1.3, is discussed in detail in Part II of this guideline, Abnormal and Emergency Operations.

The normal operations timeline in Figure 3.2 is illustrated with the pressure control points in Figure 3.5. These specific upper and lower safety limits influence the process and equipment design and operation [1] [14] [21]. Figure 3.2 will be used in subsequent chapters as each of the following transient operating modes listed in Table 1.1 are discussed:

- Shut-down for a process shutdown (Type 3, in Chapter 4)
- Start-up after a process shutdown (Type 4, in Chapter 4)
- Shut-down for a facility shutdown (Type 5, in Chapter 5)
- Start-up after a facility shutdown (Type 6, in Chapter 5)
- Shut-down for an unscheduled shutdown (Type 7, in Chapter 7)
- Start-up after an unscheduled shutdown (Type 8, in Chapter 7)
- Shut-down for an emergency shutdown (Type 9, in Chapter 8), and
- Start-up after an emergency shutdown (Type 10, in Chapter 8).

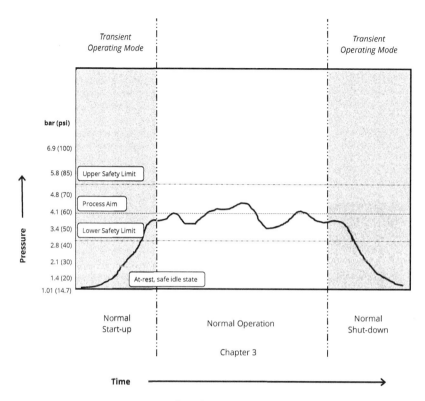

Figure 3.5 Example of pressure vessel control limits.

3.3 Procedures

Before the normal operation transient operating modes are described, this section will highlight some of the aspects which need to be written into the transient operating mode procedures, both when starting the process up and when shutting the process down. Well-written start-up and shut-down procedures help reduce mistakes. Although beyond the scope of this guideline, an organization has procedures addressing handovers between shifts, as well, to ensure that information is communicated effectively during this transition time.

Procedures for all of the transient operating modes listed in Table 1.1 may need additional, specific start-up or shut-down related steps, checklists, and decision aids to address potentially hazardous conditions that may occur during the transition, such as the following:

- Additional personal protective equipment (PPE) required during the transition
- Special handover protocols before and after scheduled projects or scheduled maintenance (Chapters 4 and 5)
- Special start-up protocols after curtailed operations (e.g., during reduced customer demand; Chapter 4)
- Special operations shut-down-related activities during weather extremes (i.e., reduce the potential for freezing when the process is not operating during the winter months; Chapter 4)
- Special start-up protocols after an emergency shut-down (especially if the end state for the process is not at its normal, safe, idle, and at-rest condition; Chapter 8), and
- Special shut-down protocols for mothballing or decommissioning equipment (Chapters 6 and 9).

An operating phase checklist noting some typical procedural steps that may need to be considered for a transient operating mode, depending on the hazards, includes the following modes [24]:

- Normal Shut-down for a Turnaround
- Start-up after a Turnaround
- Normal Shut-down for Standby Mode
- Start-Up after a Warm Shut-Down (a system put in standby mode)
- Emergency Shut-down
- Start-up after an Emergency Shut-down (ESD), and
- Initial Start-up/Commissioning.

These modes will be covered in more detail later in this Guideline.

One incident which revealed weaknesses in a routine start-up procedure occurred when fatigued heat exchanger equipment catastrophically failed during the start-up of an *adjacent* heat exchanger [25]. The start-up procedure required additional operators to assist during the start-up of the heat exchanger, resulting in many people in the vicinity of the equipment during the transition time. As both safe start-up and shut-down procedures are being developed, it is essential to ensure that the procedures do not increase the risk due to additional personnel exposure during the transition time. If the risk is increased, extra engineering and administrative controls should be a part of the routine procedure to ensure that the additional hazards are addressed. It is worth noting at this point that by relying more on administrative controls during the transition time, the likelihood increases that something might go wrong (refer to the hierarchy of controls [21, p. Figure 3.6]), and the risk is increased.

A brief description of normal shut-downs and normal start-ups after a normal shut-downs follow in Section 3.4 and Section 3.5, respectively.

3.4 Performing a normal shut-down

The normal shut-down—transient operating mode Type 1, Table 1.1— is the operating time when the process equipment is being shut-down to end a normal operation or production run, stopping all or part of the process and all or parts of its production. The operators have written plans of the steps to take the process equipment from the normal operating conditions to an idle, safe, and at-rest state. There may be special equipment-related cleaning procedures to prepare the equipment for the next production run. As was illustrated for the processes in Figure 3.2 and Figure 3.3, the normal shut-down may take some time to establish a safe, at-rest, and idle state.

A proper and safe method to shut-down or start-up a process depends on many of the plant's details, including the:

- plant type and size
- process type(s)
- processing conditions
- chemicals and materials in use
- unit operation type(s)
- equipment type(s)
- process location (indoors, outdoors)
- utility type(s)
- etc.

Each process plant has to be understood and its procedures developed in specific detail.

Normal process shut-downs will be process-specific, and for that reason, no approach described herein would apply to every process shut-down. An example approach for a safe shut-down of a continuous, pressurized process may involve the following steps:

1. Reducing the equipment processing rates and operating conditions,
2. Removing or shutting down the heating and cooling sources,
3. Stopping the feed stream(s), and
4. Reducing the equipment pressure to atmospheric pressure.

3.5 Start-up after a normal shut-down

The normal start-up—transient operating mode Type 2, Table 1.1—is the operating time when the process equipment is being started up after a normal shut-down, beginning all or parts of the process to resume production. Operators have written procedures for the steps required to take the process equipment from an idle, safe, and at-rest state to the normal operating conditions. As was illustrated for the processes in Figure 3.2 and Figure 3.3, the normal start-up may take some time to achieve the standard operating conditions.

Similar to the normal process shut-downs, normal process start-ups will be process-specific, and for that reason, no approach described herein would apply to every process start-up, either. An example approach for a safe start-up for a continuous process may involve the following steps:

1. Pressuring up the equipment (i.e., to prevent flashing of high pressure feed sources),
2. Adding external heat and/or cooling to the equipment,
3. Introducing the feed stream(s),
4. Adding heating and cooling sources to the streams, and
5. Bringing the equipment to the normal processing operating conditions.

Although each process will have specific procedures, the steps for a normal start-up procedure may include the same steps in reverse of the steps used in the process's normal shut-down procedure.

3.6 Incidents and lessons learned

Note: All incidents which occur *during* normal operations are, by definition, covered in abnormal and emergency operations (Part II of this guideline).

Before describing incidents from normal shut-downs and start-ups, it is well known that deviations from the established written start-up procedures, especially when combined with bypassing critical safety devices, can result in incidents. For example, an incident that injured four employees occurred when an automated process controlling the sequence from start to finish was interrupted and the supervisor allowed maintenance technicians troubleshooting the process to override the computer safeguards [26]. A frequent cause described in incident case studies for semi-batch processes with a high exotherm reaction was lack of agitation or delayed agitation when the rate-controlling reactant was introduced during its start-up step [27, p. 20].

For ammonia manufacturers, deviations from the start-up or shut-down operating procedures can result in reformer "burn downs" since the reforming reaction is highly endothermic. Shut-down and start-up conditions are so transient that they generate additional stresses in the radiant box due to the high temperature and pressure rate changes. One of the measures to reduce reformer tube failure includes regular visual inspection of them during start-up and shut-down, providing the operator with a clear picture of what is going on inside the furnace. Thus reliable and smooth operation of a reformer depends on the effective monitoring of catalyst tubes during these transient operating modes [28].

3.6.1   Incidents during normal shut-downs

When there are established, effective, normal shut-down operating procedures, no process safety-related incidents should occur during this transient operating mode. If a normal shut-down does cause an incident, then an incident investigation should be performed to determine why it happened [29].

3.6.2   Incidents during start-up after normal shut-downs

When there are established, effective, normal start-up operating procedures, no process safety-related incidents should occur during this transient operating mode. However, if a normal start-up does cause an incident, then an incident investigation should be performed to determine why it happened [29]. Some examples of incidents occurring during the start-up when safe, normal start-up operating procedures were modified during their execution are provided another publication [2, pp. 210-214]. Some examples of incidents occurring during the start-up of fired heaters or furnaces when lighting them during their start-up are provided other publications [20, pp. 200-203, 492] [30].

3.7 How the RBPS elements apply

All of the Risk Based Process Safety elements (RBPS) apply to effectively manage the process safety risks during normal shut-downs and start-ups (Type 1 and Type 2 transient operating modes listed in Table 1.1). In particular, the operating procedures will have been developed for the operators to take processes from their safe, idle state (the start-up), for running them under normal operating conditions (normal operations), and then taking them back to their safe, idle state (the shut-down). If these procedures have been developed and are updated as needed, the facility should not experience any incidents during these transient operating modes.

4 Process Shutdowns

4.1 Introduction

This chapter covers the transient operating modes associated with project-related shutdowns such as small Management of Change projects (MOCs) or repairs (when the equipment has maintenance or specific project-related attention that requires equipment- or process unit-related shutdowns.) It provides an overview of how inadequate handovers between groups between stages of the project, especially when starting up the equipment and processes can lead to incidents. Since the normal shut-down and normal start-up transient operating modes do not cover the additional steps required for preparing for a shutdown or for resuming operations after the changes have be made, this chapter continues with discussions on the roles of a project manager and the project team. This chapter provides a brief project life cycle overview (e.g., the stages for planning, preparing, executing, commissioning and resuming operations) with emphasis on how improving handovers can improve process safety performance during the transient operating mode.

The chapter continues with discussions on how the project-related shutdown's process shut-downs and start-ups are performed safely (Section 4.5 and Section 4.6, respectively). Then some lessons learned from incidents which occurred during these transient operating modes are Section 4.7. This chapter concludes with a discussion on the applicable RBPS elements for the planned shut-downs and start-ups afterward (Section 4.8).

4.2 The process shutdown

The two transient operating modes for a process shutdown are the shut-down beforehand (mode Type 3, Table 1.1) and the start-up afterward (Type 4, Table 1.1). A process shutdown may be referred to as an "outage," implying that the process equipment is placed in a safe state ("out of service" such as Lock Out, Tag Out {LOTO}) with the affected equipment properly prepared by operations and handed over to other groups, such as engineering, maintenance, or contractors, as needed. This differs from a "turnaround," which implies a larger, more complicated time where many process units or even an entire facility are undergoing major equipment-related work that stops production altogether (see Chapter 5). As was noted in Chapter 3, there are usually special additional handover procedures between engineering, operations, and maintenance for safe equipment ownership transfer. These handover procedures are administrative controls and are designed and implemented to reduce mistakes, reducing the special project shutdown-related risks that are associated with the process's hazardous materials and energies. The plans are designed to reduce hasty decisions to get the process back up and running quickly that many place people in harm's way.

These planned equipment outage times (the "shutdown") include product transitioning times which may require special procedures or low product demand times which affect the equipment scheduling time. However, since additional planning may be needed for small engineering projects, this chapter provides a brief overview of the types of projects requiring the different type of transition, such as when work is scheduled for authorized changes through a Management of Change (MOC) program (i.e., changes to procedures, equipment, or processes), or is scheduled for routine maintenance activities which correspond to an equipment Inspection, Testing, and Preventive Maintenance (ITPM) program [14] [21].

Since there are essential project-related planning steps used to ensure that the process equipment is prepared and ready for handovers to the group or groups, this chapter provides a brief overview in Section 4.3 of some guidance for different types of engineering projects and how to effectively manage the process safety risks during the transition times. There are several different stages in a project's life cycle, as well, which have different process safety-related risks associated with them as group handovers occur. Section 4.4 provides a brief overview of the project's life cycle phases, focusing on the times when the facility is in the transient operating mode.

In addition, a process shutdown requires several steps for effectively managing projects:

1. Planning for the projects in the shutdown,
2. Preparing the equipment for each project (if there is more than one project),
3. Executing the work safely on the isolated equipment,
4. Commissioning and confirming that the equipment is ready for the operations group, and
5. Safely starting the equipment and the process unit back up.

The two transient operating modes for a process shutdown are:

1. The shut-down mode (steps 1 and 2), and
2. The start-up mode afterwards (steps 4 and 5).

The transient operating modes before and after the process and facility shutdowns are illustrated in Figure 4.1. Each of these steps is discussed briefly in this chapter, with the process shutdown and its associated shut-down discussed in Section 4.5. Safely starting up the process afterwards is discussed in Section 4.6.

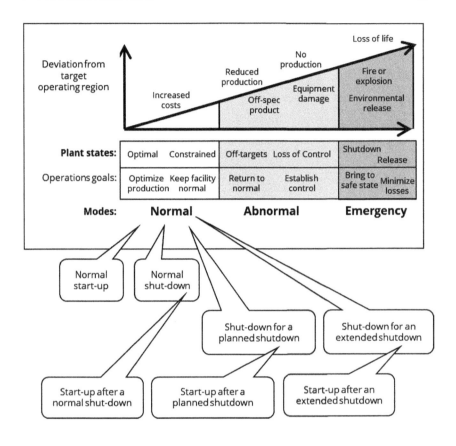

Figure 4.1 Transient operating modes associated with process and facility shutdowns.

(Adapted from [15])

4.3 Projects requiring equipment or process unit shutdowns

This section will briefly summarize how process safety risks can be addressed effectively when preparing for and executing *all types* of engineering projects. Since each project is unique, each one requires a systematic and disciplined approach to manage their process safety risks. All successful projects have structure, an execution plan and involve some form and degree of project-related, as well as process safety-related, risks that need to be understood and minimized where possible. During each stage of the project's life cycle (see Section 4.4), these risk reduction steps apply to those for both a process shutdown (this chapter) and a facility shutdown (Chapter 5).

The overview in this chapter specifically applies to an existing facility—a "brownfield"—when the equipment has been successfully commissioned and the process has been operating safely for some time. A disciplined project management approach applies to both small and large brownfield-related capital projects, which may have relatively short duration (the smaller projects) or which may occur over weeks or months (the larger projects). Projects for an undeveloped area—a "greenfield"—will be discussed in Chapter 9.

4.3.1   Types of projects

The types of projects, ranging from equipment-specific to facility-wide efforts that can be proposed, approved, budgeted, resourced, and scheduled, include the following: upgrades, debottlenecking, optimizations, and retrofits. These projects range from smaller Management of Change (MOC)-related projects to larger, resource-intensive capital projects. The distinction between the small and large projects is reflected in this guideline by the distinction between a process shutdown (e.g., with a small scope, this chapter) and a facility shutdown (e.g., with a large scope involving many groups, discussed in Chapter 5). Additional guidance with process safety-related risk evaluations for project management and project-related definitions are provided in other publications [31] [32]. Although process safety in engineering projects is covered in detail elsewhere [31], the parts of this resource that pertain to the transient operating mode, only, have been added to this guideline.

Recognizing that smaller projects are usually managed under the facility's MOC procedures, every small and large project should address the following process safety-related issues:

1. Identify the hazards;
2. Assess the risks associated with the hazards; and
3. Manage the risks to prevent and/or mitigate potential process safety, safety and occupational health, and environmental incidents.

Although many projects are initiated through the facility's engineering group, it is important that all projects—especially the smaller ones—follow the facility's MOC procedures to ensure that all hazards are properly managed. Often, smaller equipment-related projects are driven by the maintenance group since obsoleted equipment or equipment-related parts cannot be obtained anymore. Such projects should follow proper MOC-related controls to ensure that these "minor" changes to potential process safety-related issues have been properly addressed. Refer to other publications for more detailed guidance provided when managing MOCs [14] [33].

4.3.2   Managing the contractors

When the projects get larger, more groups are involved, and often contractors are required since they bring technical or construction skills expertise. Thus, contractor management becomes crucial as the project size increases. In addition, inadequate oversight of the contractor's work, especially their work's quality and verification thereof, has led to numerous incidents, described elsewhere [20, p. 597 (Index)]. More handovers between groups increases the number of opportunities for crucial information to bemissed during the communications between groups. And, when something goes horribly wrong in the project execution in the field, the adverse consequences are immediate to those working directly with the hazardous materials or energies or who happen to be located in the affected area (e.g., supervisors, engineers, operators, mechanics, and, if involved in the task, the contractors).

4.3.3   The adverse effect of inadequate handovers on the process safety risk

The number of groups involved in a project depend on the scope of the project, with more groups involved increasing the likelihood of communication miscues during handovers between the groups. Everyone needs to have the operational discipline when preparing the equipment for the project-related work, when executing the project-related work, and when starting the equipment back up. In particular, "advance planning can minimize the use of hasty decisions [3, p. 3]," especially during handovers when many different groups are involved.

Based on some of the transient operating mode incident history, this guideline will focus on the groups typically involved in the handovers that resulted in the incident. The groups typically include: management, engineering, operations, maintenance, and contractors. Other groups, such as those handling parts or equipment procurement, storage, and distribution during the project execution, may be involved as well. In all cases, these groups are responsible in one way or another, either directly or often indirectly, and may be a part of the root cause of an incident. In particular, the higher risk occurs when each group operates in its own interests and does not share critical information with the other groups. These incidents, in one way or another, had a group handover-related communication miscue or had group(s) without the operational discipline to perform their task(s) as designed during the preparations, execution, or project-related start-up phases of a process and facility shutdowns.

4.3.4   The project manager

A project manager is critical for the handovers in larger, complex projects when many groups are involved, especially since the number of groups actively involved will continuously change throughout the course of the project's life cycle. Process safety handover communications will change over the course of a project, especially between commissioning and the start-up. Depending on the size of the project, the project's groups may include staff from: design engineering; process safety; construction; quality; procurement and contracts; EHS (Environmental, Health and Safety); commissioning; and project controls. Some useful tools which help the project manager ensure that hazards and process-related risks are identified and addressed, especially during the transient operating mode, include a Project Process Safety Plan and a Hazard and Risk Register [31, p. Appendices B and C]. The essential life cycle phases involved in successfully executing a project, whether large or small, are briefly discussed in Section 4.4.

4.4 A brief project life cycle phase overview

Whether small or large, every successful project has a systematic and disciplined approach using good project management practices to ensure both the success of the project and the effective management of its process safety risks. The project life cycle is a series of phases that a project passes through from its initiation to its closure [34] [35]. Since the term "phases" is often referred to as "stages" when describing the project life cycle, this guideline will use "stages" to be consistent with the terminology used when managing the process safety risks of an engineering project [31]. (*Note:* For consistency with the transient operating mode in this guideline, "start-up" will be used for the term "startup" from this point on, as well.)

Projects that succeed have a clear framework and a clear execution plan, which is tracked and monitored at each stage of the project. These reviews occur and are managed by a project "gatekeeper" when executing a project, the project-related risks, the process safety risks, the safety and occupational health risks, the environmental risks, and the minimum regulatory compliance constraints, which should be understood and managed at all times.

As is illustrated in Figure 4.2, the stages in a large capital project life cycle are: 1) Appraise; 2) Select; 3) Define; 4) Detailed Design; 5) Construction (includes Fabrication and Installation); 6) Start-up (includes Commissioning); 7) Operation; 8) Project Phase Out; 9) Small Projects and Management of Change (MOC) efforts; 10) End of Life Project Initiation; and 11) End of Life. Each of these stages, including the distinction between the different Front-End Loading (FEL) stages 1, 2, and 3, is described in more detail elsewhere [31, p. 9].

Since many severe incidents have occurred during the construction, commissioning, and operation stages, resulting in significant impact—fatalities, injuries, environmental harm, and property damage—it is essential that the process safety hazards are understood and their risks are managed during every stage. Hazards studies and risk reviews may be required by the company during each of the life cycle stages. Construction activities include fabrication, installation, quality management, and pre-commissioning, with operational readiness activities required to prepare for commissioning, start-up, and operation. The commissioning stage concludes with the facility and its project-related documentation handed over to the operations group for normal operation. In addition, the capital project's operation stage may require additional test runs to confirm that the specific equipment or process unit performance specifications have been met *before* the handover to the operations group is completed and the operations group owns the equipment and can safely restart the process unit.

This guideline, focusing on managing the process safety risks during the transient operating mode, will discuss the shutdown preparations and subsequent start-up related aspects in Stage 6 and Stage 9, where the equipment is commissioned before the final handover to the operations group for a safe start-up. Depending upon the size of the project, commissioning and start-up may be performed by the operations group (small projects) or a separate commissioning team (e.g., large greenfield projects, see Chapter 9) with support from operations and contractor personnel as required. As a result, commissioning and start-up can involve many groups. This requires good communications between all groups involved, especially for effective and safe handovers between groups.

Whether it is a small or large brownfield-related project, there should be protocols in place to ensure that the equipment is ready and fit for duty. Start-ups for larger, greenfield-related capital projects—an "Initial Start-up"—are discussed in Chapter 9. Since this chapter focuses on the more frequent yet smaller engineering projects, there is a stronger connection to Stage 9, when the facility has already completed Stage 8 (when any larger capital projects have been phased out) and the facility is operating safely every day. The larger and more complex projects, typically requiring a long shutdown for many weeks or even months, are covered in Chapter 5.

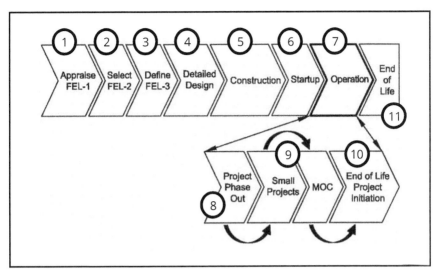

Figure 4.2 Stages in the capital project's life cycle.
(Adapted from [31, p. Figure 2.2])

Reviews—known as stage gate reviews, "cold eyes" reviews, peer reviews, project technical safety reviews, or hazards reviews, etc.—are normally conducted by an independent and experienced multi-discipline team familiar with the relevant processes and technologies. As noted earlier, larger projects require a project gatekeeper who monitors the handovers between stages, helping reduce the likelihood of incidents. Ensuring and verifying that everything is ready for each handover in advance will help reduce incidents when preparing for a shutdown and when starting the equipment and process units back up afterward.

Depending upon the scope of a project, a range of process safety activities can apply at each stage of the project, especially during the transient operating mode. Although smaller projects and management of change (MOC) modifications may combine stages or may not clearly delineate into these discrete stages, some of the activities noted apply to them. Thus, these smaller projects should still address the overall objectives of each stage to help reduce the process safety risks. The general stages used to successfully manage any size project are as follows:

1. Planning for a project's shutdown—Section 4.4.1
2. Preparing the equipment for the project's shutdown (including the handover)—Section 4.4.2
3. Safely executing the activities during the shutdown—Section 4.4.3
4. Preparing for the start-up after the project—Section 4.4.4
5. Pre-commissioning the equipment—Section 4.4.5
6. Resuming operations after the handover—Section 4.4.6

A brief overview for each stage is described next.

4.4.1   Planning for a project-related shutdown

The first stage is planning for the process equipment shut-down for the project's execution stage. Successful projects have a schedule reflecting the project's budget, the proper allocation of resources, and milestone and stage completion dates. The scope of the work should be well-defined, the expectations of and responsibilities for every group involved in the project should be clearly identified, and the deadlines for completing specific tasks should be established at the beginning. Planning tools with task-related timelines, ranging from Gantt charts to more sophisticated tools such as logic networks, are used for complex projects with multiple interactions and links between groups, resources, and the time elements.

4.4.2   Preparing the equipment for the project-related shutdown (including handovers)

Before any work is performed on the project's shutdown-related equipment, the affected equipment should be in an idle, at-rest, and safe state, with all energy sources isolated, and with the additional constraint that is *safe for people to work on it*. There may be additional procedures, checklists, and decision aids to address any additional, potentially hazardous conditions that may require special PPE when preparing the equipment for the handover. Although smaller projects can be associated with these process and equipment changes, there may be relatively fewer steps and fewer groups involved when preparing for and executing a process shutdown compared to the number of steps and group interactions required for complex, major process unit-related changes, such as revamps, expansions, or turnarounds (see Chapter 5). There will be more non-routine tasks, more administrative controls, and additional risks with the larger, more complex projects.

Incidents that occurred shut-downs when the preparation of the equipment was inadequate before being handed over to those working on the equipment (see incident discussion in Section 4.7). Handover preparations include cleaning, washing, steam and purging operations, and energy isolation (i.e., Safe Work Practices, SWP). Inadequate or inconsistent labeling of equipment in the field can also lead to incidents due to project or maintenance efforts inadvertently being executed on "live" equipment or piping [20, p. Chapter 4]. Since the handovers between groups should be clearly established, especially during pre-commissioning activities that may involve contractors and other personnel responsible for construction, installation, and commissioning, an effective "Pre-Operations Plan" should be developed before the construction stage begins (i.e., during stages 1-3, Figure 4.2).

4.4.3   Safely executing the activities during the project-related shutdown

When preparing for and executing a shutdown, there should be thorough and effective administrative controls in place and ensure effective handover communications between engineering, construction, operations, and maintenance. Communication is essential, *especially* during projects which involve equipment demolition (see Chapter 9), accessing active process lines containing hazardous materials (i.e., "hot tapping"), and simultaneous operations (SIMOPS; when nearby equipment, process units, and associated utilities may still be in operation during the project). When these routine but less frequently used specially-prepared administrative controls and associated Safe Work Practices are not properly *developed* beforehand, are not properly *understood* by those performing the work, or are not *handed over or executed* properly during the time of the equipment shutdown, incidents can (and do) occur. When incidents do occur, there can be significant harm to people, the environment, and to property (see Section 4.7 and the Marsh data [36]).

4.4.4   Preparing for the start-up after project completion

As noted earlier in this chapter, a gatekeeper review is essential when preparing for shutdowns. This review should focus on verifying and confirming the work performed during construction (Stage 5) is ready for handover to the groups managing the start-up (Stage 6). After the shutdown is over, when Operations group is ready to receive the equipment and start its production, another gatekeeper review confirms and verifies that the equipment is ready to be handed over. Without a formal Operational Readiness Review (ORR) for larger projects, the likelihood of incidents increases when the process is started back up. For smaller engineering projects, the gatekeeping steps are required, as well, and are typically designed into the facility's

MOC system which includes a robust Pre-Startup Safety Review (PSSR). Details to a series of process safety-related questions which can be posed by the gatekeeper review during each stage of the life cycle, including the start-up, are provided elsewhere [31, p. Appendix G].

4.4.5 Pre-commissioning the equipment

Depending on the scope of the project, the following pre-commissioning activities may be needed to address the newly installed or modified equipment (Adapted from [31, p. Table 2.2]):

- Identifying and documenting all process safety information received from the suppliers and vendors;
- Vessel and pipe
  - o Flushing and chemical cleaning
  - o Dewatering and drying or blowing (e.g., air, nitrogen, vacuum, methanol or glycol swabbing)
  - o Pressure testing (both hydrostatic and pneumatic)
  - o Leak testing;
- Internal inspections;
- Design conformity checks (visual and review test certificate reviews);
- Control system checks;
- Instrument loop checks and calibrations;
- Safety system functional tests and validations (e.g., interlocks, emergency shutdowns);
- Safety Instrumented System (SIS), Emergency Shut-down Systems (ESS), Safety Shutdown System (SSD), or safety interlock system functional tests and validations;
- Electrical continuity and motor rotation checks (including electrical isolation work permits);
- Equipment static and de-energized tests;
- Equipment (machinery) lubrication, cold-alignment, and guarding; and
- Pipeline gauging to identify buckling, dents, and other damage.

Note that some of these pre-commissioning activities will require access to utility supplies, such as water or compressed air/nitrogen, and others will require electricity to energize the equipment being tested (e.g., when checking the rotation direction of electrical motors). Every one of these pre-commissioning activities may adversely affect the process safety risks during the start-up if they have not been evaluated and performed correctly.

Larger projects may have pre-commissioning activities within the construction site that require additional safeguards and management. Some of these safeguards include, but are not limited to [31, p. 167]:

- Barriers to exclude personnel from specific areas (e.g., during hydrostatic pressure testing);
- 'Blanket work permit' removal afterwards;
- Communicating the changed status to the workforce.

One of the pre-commissioning activities often performed is called "punch-listing," which helps identify, record, and correct equipment that may have been damaged, as well as incorrectly or incompletely fabricated or installed during construction. Items that can be identified can be categorized into three general topics as follows (Adapted from [31, p. Table 7.3]):

1. To be corrected *before* commissioning - Items that do not meet safety and design standards, or which prevent commissioning (e.g., missing or damaged items and incorrectly fitted items, such as loose bolts or wires)
2. To be corrected *prior to handover* to the operations group - Items that may be completed during commissioning (e.g., missing signs and labels or long bolts in flanges)
3. To be *scheduled in agreement* with operations - Items that are cosmetic and do not prevent start-up (e.g., painting and non-critical insulation).

Again, the handovers between groups should be clearly established, especially since these pre-commissioning activities may involve contractors and other personnel responsible for construction, installation, and commissioning. For this reason, an effective "Pre-Operations Plan" will have been developed before the construction stage begins (i.e., during stages 1-3, Figure 4.2). These key activities for these earlier stages are described in more detail elsewhere [31, p. 174].

4.4.6   Resuming operations after the handover

When restarting the process equipment after a project-related or maintenance-related shutdown, the operations group depends on an effective handover from the group or groups working on the equipment. The special administrative procedures used to prepare the equipment for the duration of the shutdown, if any, should be addressed so that the equipment is returned to its normal, safe, at-rest and idle state before start-up. This includes, for example, removing all blinds that had been added to isolate the equipment from other parts of the process equipment during the project or maintenance work. When all the special equipment preparations have been reversed, the operations group then can use their normal start-up procedures to restart the process safely (see Chapter 3).

Often the final step used to close-out a larger project is to review the project's successes and challenges, including a lessons learned review to improve the facility's project management system performance for future projects. Any issues which had to be addressed and solved during the project, including those during shut-down and start-up, are lessons that capture the knowledge gained from the project's experiences. At this point, the project team officially phases out, the facility is handed over completely to the operations group, and the project is closed.

This concludes the overview section of the project life cycle. In summary, the basic stages for every project are as follows: planning for a project's shutdown; preparing the equipment for the project's shutdown; safely executing the activities during the shutdown; preparing for the start-up after the project; resuming operations after the handover; and, for larger projects, reviewing the project's successes and challenges. The next section discusses the two transient operating modes associated with a process shutdown: preparing the shut-down for a process shutdown (Section 4.4) and resuming operations with a start-up afterwards (Section 4.5).

4.5 Preparing for planned project-related shutdowns

The shut-down for a process shutdown—transient operating mode Type 3, Table 1.1—is the operating time that may require additional, often non-routine procedures that are used during or immediately after the normal shut-down mode. This additional time is illustrated in Figure 4.3. The process shutdown may include low demand periods, smaller projects performed through a Management of Change (MOC) program (e.g., changes to procedures, processing conditions, processing equipment, control systems, or utilities), and routine maintenance activities (e.g., through an Inspection, Testing, and Preventive Maintenance (ITPM) program). Thus, the production personnel may have additional, specially written plans for the different steps required to prepare the process equipment for a planned equipment-related activity.

Special operations-to-maintenance handover procedures and checklists may be required, including lists of the steps required for cleaning or isolating the equipment before maintenance can work on the equipment. These additional administrative controls, including Safe Work Practices (SWP), help reduce the likelihood of personnel exposure to any hazardous materials and energies when they are adequately written, understood, and followed. Additional guidance is provided for designing effective SWP to help reduce the risks to personnel at these times [14] [37]. The shut-down time for a process shutdown may typically take more time than the normal shut-down time due to these additional administrative controls.

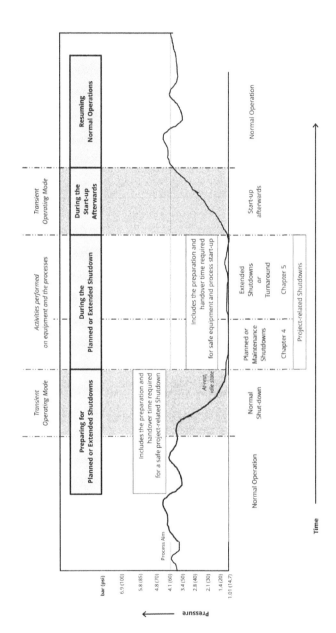

Figure 4.3 Example timeline for a process or facility shutdown.

Since these additional procedures may not be performed very often, it is essential that everyone involved in a shut-down for a process shutdown understands what the different steps are, has the operational discipline to follow these steps, and can quickly recognize and respond properly when things are not going as planned. Many of the US CSB incidents show that unreviewed and unapproved changes in the field have led to severe incidents when executing nonroutine shut-downs for project-related activities. When people do not adequately assess and address the hazards, or when people make changes to the established, approved plan without understanding how their changes increased the process safety risk, consequences can include severe injuries, fatalities, environmental harm, and property damage. Due to the harm which occurs, project plans should be thoroughly reviewed and approved by every group involved in the planning and execution the project-related shutdown, especially those in operations, maintenance, and engineering.

Effective handover protocols and systems should be in place, as well, to ensure that those working on the equipment know what hazards have been or have **NOT** been addressed before the work commences. There may be special clean-out or isolation procedures required for the project that are not done during a normal shut-down transition. There may be special hazards that are introduced to make the equipment safer to work on that should be carefully monitored during the shutdown-related work, such as displacing toxic gases or flammable vapors with an inert, asphyxiating gas (i.e., Nitrogen). A robust Management of Change (MOC) system can help ensure that everything is ready, that all the equipment is prepared and in a known state by everyone before beginning the scheduled work [33]. Ensuring and verifying that everything is ready will help reduce the miscues that have led to significant incidents.

4.6 Start-up after planned project-related shutdowns

The start-up after a process shutdown—transient operating mode Type 4, Table 1.1—is the operating time that may require non-routine procedures, in addition to the normal start-up procedures, before resuming all or parts of the process and all or parts of its production. If other groups were involved in the shutdown, such as engineering and maintenance, special handover procedures should be in place, including performing an Operations Readiness Review (ORR) for larger projects, which may include a Pre-Startup Safety Review (PSSR) —an integral part of the facility's smaller project's MOC system—before resuming operations. A start-up after a project-related shutdown typically takes more time than a normal start-up due to the additional procedures involved, including the time for effectively completing the handovers inherent in an operational readiness review.

As noted earlier, many of the US CSB incidents show that inadequate handovers to the operations group after a project-related shutdown have led to severe incidents when starting the equipment back up [5]. Other publications have discussed the risks associated with these times, as well [3] [4] [9]. Technical failures influenced by inadequate maintenance have contributed to incidents during start-ups and shut-downs [6, p. 346]. Due to the harm which may occur, project-ending start-ups should be clearly understood by the operation group, all of whom should have been adequately trained to operate the equipment with updated or new procedures, as needed. To help reduce the miscues which have led to incidents when starting the equipment and process units back up, the facility can build into its systems project stage handover-related reviews. These reviews verify and confirm that the equipment installed in the field has been built or modified per the design standards (during the construction stage).

Inadequately managed changes, such as installing the wrong materials of construction during the shutdown period, can lead (and has led) to unexpected and significant incidents (sometimes *years* later). Examples of different materials of construction related incidents have been reported elsewhere [20, p. Chapters 16 and 28]. These include using a titanium flange on a line carrying dry chlorine, using carbon steel instead of stainless steel or suitable alloy, adding acid to an aluminum tanker assuming it was made out of similar-looking stainless steel, and using off-specification bolts. These materials could have been specified incorrectly during the design, been purchased incorrectly because they were less expensive, or installed incorrectly since they were not located in dedicated warehouse space, or simply installed incorrectly since they looked similar. For these reasons, a facility should verify that the correct design, purchase, acquisition, and installation of the component or equipment occurs during each group handover before start-up of the process.

Effective handovers can occur when there is a review between the Construction Stage 5, the "Startup" Stage 6 [in this guideline, Start-up is used], and the Operations Stage 7 before resuming operations (Figure 4.2). The first set of topics that a gatekeeper could cover with the review team, for the handover from the construction stage to the commissioning and start-up stage, is listed in Table 4.1. The next set covered, before the operations group starts the equipment back up (Stage 7), is the handover from the commissioning and start-up stage to the operations stage (Table 4.2). These tables present typical issues that a review team may address as a part of their protocol when covering each stage, thus helping reduce the likelihood of an incident upon restart.

Table 4.1 Handover topics between the construction and the commissioning/start-up stages.

(Adapted from [31, p. Appendix G])

Stage 5	Project Life Cycle Stage Gate Review From the Construction Stage to Commissioning and Start-up Stages
	Handover Topics
5.01	Confirm that construction workforce training, competency, and performance assurance arrangements are adequate and being implemented.
5.02	Confirm that a construction Process Safety management system was adequate and was being implemented.
5.03	Confirm that owner, contractors and vendors have clarity in regard to their scope and responsibilities for the mechanical completion, and that the construction team have a robust process to manage all interfaces.
5.04	Confirm that asset integrity management (AIM) processes including quality management are sufficient to deliver the design intent and facility integrity.
5.05	Confirm that change management was being applied.
5.06	Confirm that project plans for pre-commissioning, commissioning, and pre-start-up are adequate.
5.07	Confirm that progress on Operations training and development (or update) of operating procedures was adequate.
5.08	Confirm that the Operations Team was involved as necessary in preparation for pre-commissioning and commissioning activities.
5.09	Confirm that plans for a site Process Safety management system and procedures are adequate.
5.10	Confirm that a document management system has been implemented and was performing as expected.

Table 4.2 Handover topics between the commissioning/start-up and the operations stages.

(Adapted from [31, p. Appendix G])

Stage 6	Project Life Cycle Stage Gate Review From the Commissioning and start-up ("Pre-start-up") to Operations
	Handover Topics
6.01	Confirm that pre-commissioning has been satisfactorily completed and the facilities are ready for commissioning.
6.02	Verify that project and/or site is implementing a comprehensive process to confirm preparedness (e.g. Operational Readiness Review) and obtain approvals for start-up.
6.03	Confirm that integrity of the design has been maintained, and deviations from design have been satisfactorily addressed and will not compromise Process Safety performance.
6.04	Confirm that the commissioning, start-up and operations teams are adequately trained, equipped, and competent and that all necessary procedures are available.
6.05	Confirm that the owner / site has made adequate preparations for start-up.
6.06	Confirm that emergency response arrangements and procedures have been established.

4.7 Incidents and lessons learned

Many project-related incidents occurred when equipment was not adequately prepared for the project-related shutdowns or special maintenance-related efforts. This section will provide details on a few incidents, however the reader is encouraged to review additional resources with guidance that can be used to prevent them, such as the following:

- Guidance for cleaning, washing, steam and purging operations in preparation for maintenance [2, pp. 149-161] [38].

- Guidance to help prevent maintenance-related incidents during the shutdown, and some examples of incidents occurring during maintenance due to inadequate preparations after the shut-down [20, pp. 12-17].
- Guidance for removing the hazards from equipment before the shutdown activities begin [20, pp. 17-23].

Addition maintenance-related guidance is provided in more detail, as well [2, p. Chapter 7] [20, p. Chapters 1 and 23]. Some specific incidents during the shut-down and start-up transition times are noted next. The incident summary is provided in the Appendix.

4.7.1   Incidents during shut-downs for planned project-related shutdowns

C4.7.1-1 – Delaware City Refinery Company (DCRC) Equipment Preparation [39]

Incident Year: 2015

Cause of incident occurring during the preparation time: Leak of hydrocarbons through a "closed" single block valve while personnel were preparing equipment for maintenance work by de-inventorying and draining vessels located between two isolation points.

Incident impact: Release of hydrocarbons into the sewer, which ignited and caused a flash fire that injured an operator with second- and third-degree burns

Risk management system weaknesses:

LL1) Operational tasks for preparation for maintenance can be uncommon and non-routine. The hazards and risks should be assessed when preparing the equipment, including establishing clear procedures to perform the task safely. Any changes to the pre-plan during equipment preparation should follow a structured change management protocol.

Relevant RBPS Elements:

- Hazard Identification and Risk Analysis
- Operating Procedures
- Management of Change

LL2) Equipment isolation for de-inventorying or decontamination activities prior to maintenance should use a double-block-and-bleed valve design and not drain contents directly into sewers.

Relevant RBPS Element:

- Process Knowledge Management

4.7.2   Incidents during start-up after planned project-related shutdowns

C4.7.2-1 – Maintenance Replacement [2, pp. 50-52]

Incident Year: Not noted; published incident in 2015

Cause of incident: Similar, available, and not-per-design-specifications check valve installed during routine seal maintenance on a large, offline pump

Incident impact: Release of 18,000–23,000 kg (20-25 tons) of flammable light hydrocarbon which ignited and burned for more than 5 days, resulting in a costly three-month business interruption

Risk management system weaknesses:

LL1) The replacement valve was the only check valve in the warehouse that could be fitted in the 0.3 m (12 inch) 21 bar (300 psi) discharge line of the large petroleum pump. Although the replacement check valve matched the piping's diameter and pressure rating, it required a small spool piece and additional gasket to bridge the gap in the piping. The failed gasket material was rubber suitable for fire water service, not hot hydrocarbon service. Recommendations from the incident review included upgrades to the site's change program, especially "small changes," and implementation of its recently developed materials verification program.

Relevant RBPS Elements

- Hazard Identification and Risk Analysis
- Asset Integrity and Reliability
- Management of Change
- Operational Readiness

4.8 How the RBPS elements apply

All of the Risk Based Process Safety elements (RBPS) apply when setting up a process safety and risk management program to effectively manage the process safety risks during the shut-downs for process shutdowns and the start-ups afterwards (the Type 3 and Type 4 transient operating modes listed in Table 1.1). Additional guidance for applying and auditing these RBPS elements for an effective overall process safety and risk management program is provided in other resources [14] [40].

5 Facility Shutdowns

5.1 Introduction

This chapter discusses the transient operating modes associated with a facility shutdown: the shut-down for a facility shutdown (mode Type 5, Table 1.1) and the start-up afterward (Type 6, Table 1.1). The considerations and types of larger projects requiring a process unit or facility project-related shutdown are described in Section 5.3. This chapter then provides a discussion on preparing for a facility project-related shutdown (Section 5.4), starting up after a shutdown (Section 5.5), and provides an incident that occurred during these transient operating modes with lessons learned. This chapter concludes with a discussion on the applicable RBPS elements for the facility shut-downs and start-ups afterwards (Section 5.7).

5.2 The facility shutdown

Larger capital projects that involve shutting down the equipment associated with an entire process unit, an entire facility, or even multiple, interconnected production facilities are designated as a facility shutdown in this guideline. These projects are complex, involve many different groups, and—distinguishing them from the smaller, planned projects discussed in Chapter 4—have an extended timeline for the work, covering weeks or even months of both the preparation time beforehand and the execution time during the project. Often the longer, larger projects that stop operations for the long period are called a turnaround or an outage, as well. The facility project timeline was shown in Figure 4.3. These projects are intense and often stressful times due to their scope (or scopes), and the complex demands can affect the cross-function handovers—inadequate handovers can lead to incidents. Undue stress on the maintenance group to "stay on schedule" may also be attributed to equipment-related issues which arise during start-ups.

Although companies may use different terms for these complex, production-stopping projects, one common industry term used is the "turnaround." Typically, turnarounds are defined for scheduled shutdowns or other large maintenance activities [41]. The handovers between groups can be effectively managed through a cross-functional Project Management Team (PMT), which ensures that:

- The operations group has
  o Properly prepared the equipment for the specific project-related work during shut-down,
  o Clearly established the condition of the equipment before the handover, such as having:
    - Removed hazardous materials, and
    - Cleaned the equipment
- All handovers between the different groups associated with the project, such as engineering, maintenance, or contractors, are managed during all stages, *and*
- The equipment is safe and fit for use before being returned to the operations group for start-up.

As was noted in Chapter 3, Section 3.3, there are usually special, additional handover procedures from engineering and operations to maintenance for safe equipment ownership transfer. These administrative controls are designed and implemented to reduce mistakes, minimizing the special project shutdown-related risks which may be associated with the process's hazardous materials and energies. The essential project-related planning steps were discussed in Chapter 4, Section 4.2, and the typical project life cycle was discussed in Chapter 4, Section 4.4.

5.3 Projects requiring a process unit or facility project-related shutdown

This section describes brownfield project-related considerations that may apply to facility shutdowns *in addition* to those discussed for a process shutdown (see Chapter 4, Section 4.3). (Greenfield projects will be discussed in Chapter 9.) Facility shutdowns also rely on a robust protocol for effectively managing all the steps in the project: 1) planning, 2) preparing the equipment, 3) executing the work, 4) commissioning, and then 5) safely starting back up. These steps were described in Chapter 4, Section 4.4., with the two transient operating modes for a facility shutdown discussed in this chapter:

1. The shut-down beforehand (steps 1 and 2 listed above; Type 5, Table 1.1), and
2. The start-up afterwards (steps 4 and 5 listed above; Type 6, Table 1.1).

It is essential for larger projects to use a systematic and disciplined approach to manage the project's process safety risks using these project life cycle stages. Although these stages apply to both small and large projects, a facility shutdown *should have* a capable project manager who can effectively lead the complex project's cross-functional Project Management Team (PMT) through each of the project's stages, *especially* during handovers between groups (see Chapter 4, Section 4.3.3). The effective project manager can help the company achieve success for both the project-related scope (on time and on budget), *as well as* the process safety risks (no incidents). This section discusses some additional considerations which may apply to a complex project: additional brownfield-related project considerations (Section 5.3.1), retrofit and expansion projects (Section 5.3.2), control system upgrades associated with the project (Section

5.3.3), mothballing projects (Section 5.3.4), and managing contractors during the project (Section 5.3.5).

5.3.1   Additional brownfield-related project considerations

As discussed in Chapter 4, every project should address the following process safety-related issues: 1) identify the hazards; 2) assess the risks associated with the hazards; and 3) manage the risks to prevent and/or mitigate potential process safety, personnel safety and occupational health, and environmental incidents—especially larger brownfield-related projects. The larger projects depend and rely on more administrative and procedural controls, especially during equipment handovers between groups. Relying more on administrative controls during the transition time increases the likelihood that something might go wrong (refer to the hierarchy of controls [21, p. Figure 3.6]). A weak process safety culture and leadership foundation, in combination with weak operational discipline foundation, can result in incidents, especially during facility shutdowns.

5.3.2   Retrofit and expansion projects

Retrofit, expansion, debottlenecking, upgrade, optimization, and revamp projects share many of the same challenges. Although these projects may only replace or update an existing facility while maintaining production at the existing capacity, other retrofit and revamp projects may be required to handle issues related to aging facilities [22]. These projects can introduce new or greater process safety-related hazards that should be understood and managed [21, pp. 175-177] [42, pp. 65-73]. Again, a greater reliance on procedures (administrative controls) needs leadership support to maintain a robust process safety and risk management system as well as strong operational discipline from everyone at all levels in the organization to effectively manage the procedures (refer to the hierarchy of controls

[21, p. Figure 3.6]). Inadequate handovers during these project-related shut-downs and start-ups have led to incidents, as well.

5.3.3 Control system upgrades associated with the project

Larger projects on older facilities often include Basic Process Control System (BPCS) upgrades due to rapidly and continually changing computer technologies. The manufactures of the original equipment may not exist anymore, parts for the existing hardware may no longer be available, or the manufacturer may no longer support the original version of the control system. Improvement in how to better address the design and engineering of human factors-related issues, in particular, the Human-Machine Interface (HMI), often become part of the larger project's scope, as well.

Other issues which may need to be addressed include:

- Input/Output (I/O) equipment, such as detectors and transmitters in the field;
- Interface connections between the existing field equipment and wiring;
- Network infrastructure and connectivity;
- Ancillary systems (e.g., power, including an Uninterruptible Power Supply or Source (UPS), space requirements, and
- Out-of-date documentation for the existing system.

Some of these issues may involve significant resources, especially when larger projects involve many personnel from the different groups involved: hardware and software designers, human factors experts, control system experts, process safety experts, field engineers, operators, mechanics, electricians, technicians, contractors, and trainers. Inadequate handovers between groups during these project shut-downs and start-ups can lead to incidents.

Note that these control system projects are likely to have significant training requirements for both engineering and operations personnel who will have to use the new systems every day. Control system upgrades tend to add anxiety about starting up the process on the new system. The situation tends to be worse for individuals not savy enough to adopt to the newer technology. Thus, it is critical that operators and engineers receive simulator-based training rigorous enough to create confidence about using the new capabilities.

5.3.4 Mothballing projects

Mothballed processes and equipment are temporarily shutdown for an unknown period of time, requiring some form of preservation during the shutdown period. Special procedures and checklists may be required, including lists of the steps required for cleaning or isolating the equipment before idling the period for an unknown period. The main challenge for mothballed equipment is preventing deterioration so that the facility may be safely put back into production. Sometimes efforts on the larger capital projects currently underway should be stopped temporarily with the project completed at a later date. Such costly delays may be due to issues beyond the company's control, including unexpected, sudden, or adverse economic conditions, or significant technology, construction, or installation issues which were not anticipated in the original scope (and, hence, were not budgeted when the capital project was originally approved).

The type of preservation techniques needed will depend on the stage that the mothballed equipment is in when the mothballing will occur. Preservation procedures will differ if the equipment is still in construction stage, if the equipment has been received and is staged for installation, if the equipment is partially or fully installed in the field, or if, for existing processes, the equipment has been operating for some time and needs to be cleaned before being mothballed. Depending on the equipment metallurgies and location (especially outdoors), the following preservation techniques may need to be considered [31, p. 25]:

- Periodically turning or rotating motors;
- Capping of open piping connections to partially installed equipment, vents to atmosphere, or flares;
- Maintaining nitrogen blankets;
- Coating or filling machinery with oil, and
- Using desiccants or biocides if necessary.

In all cases, a multi-discipline project team, including process safety and asset integrity experts familiar with the equipment's design, should determine the appropriate preservation approaches. Further information and guidance on asset integrity of mothballed facilities is available elsewhere [23] [43]. Inadequate preparation and handover during these project shut-downs and start-ups have led to incidents.

5.3.5 Effective contractor management during the project

Contractors are more likely to be involved in facility shutdowns due to their technical expertise, construction-related skills, and broader experience base. For this reason, contractor management becomes much more crucial since there are more opportunities for essential information to be missed during the handover communications between groups. The contractors and their sub-contractors have a clear understanding of all process safety expectations, the safe work practices, and most importantly, they have the operational discipline to adhere to them. A case study showing the successful application of a successful contractor-managing approach for major, complex facility shutdowns is provided in Section 5.6 [44]. Further information and guidance on contractor management is available elsewhere [14] [31].

When managing a facility shutdown, complex project-related issues that can adversely affect process safety can and do arise. The next section will cover the two facility shutdown transient operating modes: preparing for a facility project-related shutdown (Section 5.4) and starting up afterward (Section 5.5).

5.4 Preparing for a facility project-related shutdown

The shut-down for a facility shutdown—transient operating mode Type 5, Table 1.1—is another transient operating time that requires non-routine procedures in addition to the normal shut-down procedures when stopping the process equipment. Facility shutdown projects are complex, involve many different groups, and have work-related timelines for weeks or even months, including the preparation time beforehand and the execution time during the project. The additional time for a facility shutdown was illustrated in Figure 4.3.

Personnel in the operations and maintenance groups typically have additional, sometimes specifically, project-related procedures for the steps used to prepare the process equipment for the extended period. These additional administrative controls, including Safe Work Practices (SWP), help reduce the likelihood of personnel exposure to any hazardous materials and energies when they are adequately written, understood, and followed [14] [21] [37]. Facility shutdowns typically take more time than a process shutdown due to these additional administrative controls for shutting the equipment down, preparing the equipment, working on or maintain the preservation procedures, and then preparing the equipment for handover back to the operating group.

Since these additional procedures may not be performed very often, it is essential that everyone involved in a facility shutdown understands what the different steps are, has the operational discipline to follow these steps, and can quickly recognize and respond properly when things are not going as planned. When hazards have not been adequately assessed and addressed, or when changes to the established, approved plan have been made without understanding how their perceived solutions increased the process safety risk, severe injuries, fatalities, environmental harm, and property damage occur. Project plans should be thoroughly reviewed and approved by every group involved in the planning and safe execution of the shutdown, including those in operations, maintenance, and engineering.

Effective handover protocols and systems should be in place, as well, to ensure that those working on the equipment know what hazards have been or have **NOT** been addressed before the work commences. The equipment may need special clean-out or isolation procedures, or new hazards may be introduced to make the equipment safer to work on during the project (e.g., displacing toxic gases or flammable vapors with nitrogen or other inert gas). A robust project-management system will help ensure that everything is ready, that all the equipment is prepared and in a known state by everyone before beginning the scheduled work [14] [33]. Ensuring and verifying that everything is ready will help reduce the handover-related miscues which have led to significant incidents.

5.5 Start-up after a facility project-related shutdown

The start-up after a facility shutdown—transient operating mode Type 6, Table 1.1—is the operating time that may require non-routine procedures, in addition to the normal start-up procedures, when planning for and then resuming operations. If other groups were involved in the facility shutdown, such as engineering, maintenance, and contractors, special handover procedures should be in place, including performing an Operations Readiness Review (ORR) for larger projects, which may include a Pre-Startup Safety Review (PSSR)—an integral part of the facility's smaller project's MOC system—before resuming operations. Start-ups after a facility shutdown typically take more time than a normal or planned project start-up due to the additional project preparation and shutdown-related procedures involved, including the time for a thorough Operational Readiness Review.

Good operational discipline, combined with effective process safety leadership and systems, is needed to decrease the likelihood of severe incidents [21]. In addition, both the Marsh and US CSB incidents show that severe incidents have occurred when starting the equipment back up after a planned or a facility shutdown due, in part, to inadequate handovers of processes from different groups to the operations group [3] [4] [9] [36]. The hazards remaining (if any) during end-of-project-related start-ups should be clearly understood by the operators. If new training is essential for the safe operation of the new or upgraded equipment, the operators should be adequately trained to operate the equipment with updated or new procedures before the start-up commences.

5.5.1   Pre-commissioning the equipment

As was discussed in Chapter 4, Section 4.4, management can build specific facility shutdown handover-related reviews into its systems at each stage in the project's life cycle to verify and confirm that the equipment installed in the field has been built or modified per the design standards and is ready to be handed back to the operations group. These include reviews for pre-commissioning the equipment and using punch-lists. An example of potential handover review topics was listed in Chapter 4, Table 4.1 (includes contractors, the group(s) preparing for the start-up) and Table 4.2 (includes the commissioning and pre-start group(s) before the operations group starts back up). These tables represent typical issues that a review team may identify to address and verify the work that should be completed at each stage in the project's life cycle. In addition, if a facility shutdown contains specially-protected equipment (e.g., it has been mothballed during the shutdown period since it was not being worked on), the measures implemented to preserve the equipment should be addressed when pre-commissioning the equipment (Section 5.5.2).

5.5.2 Recommissioning mothballed equipment

The recommissioning of mothballed equipment will depend upon how long the equipment has been mothballed and how well the prevention techniques, if any, were maintained (Section 5.3.4). In all cases, a multi-discipline project team should be assigned to inspect and test the mothballed equipment to determine its integrity. The preservation approaches used will need to be validated (e.g., rotating the motors) or reversed (e.g., coating the equipment with oil), as needed. For mothballed equipment that had been in operation and was shutdown, the recommissioning team should perform an Operations Readiness Review (ORR) before the operations group restarts the equipment and its associated processes [14].

5.6 Incidents and lessons learned

Details of some facility shutdown-related incidents are included in this section. The incident summary is provided in the Appendix.

5.6.1   Case study with no process safety-related incidents

C5.6.1-1 – Dolphin Energy Limited (DEL) [44]

Project's Years: 2003-2010

Cause of the process shut-down: A major facility shutdown
The project covered seven years from concept development and preliminary engineering to full operations, involving one onshore sour gas processing plant and two offshore platforms, each with twelve wells and a separate 80 km (50 mile), 0.9 m (36 inch) underwater supply line from the platform to the onshore sour gas plant. The project also included the 364 km (226 mile) underwater pipeline from the onshore plant export gas plant to receiving facilities through an onshore distribution pipeline/network.

There were two major project phases associated with the commissioning and safe start-up of the equipment (Table 5.1). Phase one ended with a Warranty Shutdown in 2009 that lasted thirty days; phase two, a Maintenance Shutdown in 2010, lasted eleven days. At the project's peak manpower times, there were 986 people in the first phase and 272 people in the second phase. The number of work permits exceeded 1,000 for phase one and 300 permits for phase two, with 520MM hours contractor time logged for phase one and 79MM hours logged for phase two.

The first phase involved coordinating the shutdown-related efforts with the units that were not a part of the shutdown. These Simultaneous Operations (SIMOPS) were safely executed due to rigorous communications protocols established through the project's Operations Shutdown Team (Ops SD Team).

Tracking of incidents during the shut-down (or start-up):
There were no process safety incidents, no lost-time injuries, and no medical treatment cases during the seven-year effort (Table 5.2).

Risk management system strengths:
LS1) Shutdown preparation and planning (including handover procedures)
The Ops SD Team began before the effort almost nine months beforehand due to the combination of long lead times for the chemicals and catalysts and the extensive Piping and Instrumentation Diagram (P&ID) mark-ups required for the project. These preparatory efforts included developing a comprehensive list of the critical isolation blinds once the full scope of the modified or inspected equipment was established.

Table 5.1 Scope of work for the Dolphin Energy Limited shutdowns.

	Warranty Shutdown Feb 2009	**Maintenance Shutdown Feb 2010**
Duration (days)	30	11
Units shutdown	All of Stream 1	Multiple Units
No of vessels	126 (+ piping modifications)	20 (+ piping modifications)
Contractor mhrs	520,000	79,000
Peak manpower	986	272
No of blinds	1,400	390
No of permits	>1,000	300

Table 5.2 Safety summary for the Dolphin Energy Limited shutdowns.

	Warranty Shutdown Feb 2009	Maintenance Shutdown Feb 2010
No of LTI	0	0
No of MTC/FAC	0/1	0/1
No of NM reports	6	5
No of TBT	1,808	80
No of training courses	1,103	250
No of TRAs	150	30
No of safety awards	5	2

Note for Table 5.2: Lost Time Injury (LTI); Medical Treatment Case (MTC); First Aid Case (FAC); Near Miss (NM) incidents; Toolbox Talk (TBT) that were translated into several languages for the diverse workforce; and Total Risk Assessments (TRAs)

The Ops SD Team, using DEL's required Mechanical Isolation Procedures, developed Work Method Statements (WMS), typically one page, which stated the following [44, p. 4]:

• the purpose for taking the equipment out of service;
• an outline of the work to be performed;
• the Scope of Work (SOW) for Operations/Utilities groups; and
• a clear identification of the equipment to be handed over from Operations/Utilities to the Ops SD Team and then to Contractor.

Upon the completion of the work, the handover sequence was reversed.

Relevant RBPS Elements [14] [40] [31]

• Contractor Management
• Hazard Identification and Risk Analysis

5.6.2   Incidents during shut-downs for a facility project-related shutdown

C5.6.2-1 – Husky Superior Refinery [45]

Incident Year: 2018

Cause of incident occurring during the shut-down preparations: Air flowed forward into the hydrocarbon side of the Fluid Catalytic Cracking Unit (FCCU) through the Spent Catalyst Side Valve (SCSV) failing open due to erosion.
Incident impact: Ignition of the hydrocarbons in the FCCU; explosion damage nearby equipment and caused fires; thirty-six injuries onsite; evacuation of portion of community surrounding facility.

Risk management system weaknesses:

LL1) Lack of an effective Inspection, Testing, and Preventive Maintenance (ITPM) program to detect failure of Spent Catalyst Slide Valve (SCSV) due to erosion in its orifice port.

- Asset Integrity and Reliability

5.6.3   Incidents during start-up after a facility project-related shutdown

C5.6.3-1   BP Texas City Refinery Explosion
[8, pp. 60-67] [20, pp. 414-416] [21, pp. 385, Indexed] [46] [47] [48] [49]

Incident Year: 2005

Cause of the process shut-down: Major equipment turnaround

Cause of incident occurring during the start-up: Column overfilled, released hydrocarbons into blowdown drum open to the atmosphere. The blowdown drum overflowed, releasing the flammable hydrocarbons into the process area. The hydrocarbons pooled and quickly formed a vapor cloud that ignited and exploded.

Incident impact: There were fifteen fatalities, 180 injuries, a community shelter-in-place order affecting 43,000 people, and significant property damage, including residences almost 1.2 km (0.75 miles) away.

Risk management system weaknesses:

LL1) At the time of the incident, leadership at every level had emphasized personal safety but had not emphasized, reviewed, audited, or measured its process safety programs as it managed its process hazards and risks. The leadership at the Texas City refinery at the time of the incident had not established the essential positive, trusting, and open environment with its workforce, as well. In addition, at the time of the incident, leadership had not provided sufficient training or resources (including contractors), especially with process safety competencies and capabilities; had allowed high overtime rates for operations and maintenance personnel; ensured that personnel effectively follow the safe work procedures (i.e., permit to work; job safety analyses), and did not ensure that proper start-up procedures were in place—and being used—before operations resumed.

Relevant RBPS Elements
- Process Safety Culture
- Process Safety Competency
- Workforce Involvement
- Operating Procedures
- Safe Work Practices
- Contractor Management
- Training and Performance Assurance
- Operational Readiness
- Measurement and Metrics

- Auditing
- Management Review and Continuous Improvement

LL2) The corporate safety management system did not establish the operational discipline—a robust conduct of operations—to ensure timely compliance with the BP's internal process safety standards and programs, timely implementation of external good engineering practices, thorough reviews for identifying the hazards, effectively assessing for process safety risks, and maintaining process safety knowledge.

Relevant RBPS Elements
- Compliance with Standards
- Process Knowledge Management
- Hazards Identification and Risk Analysis
- Conduct of Operations

LL3) Critical equipment and their safeguards had not been identified and did not have scheduled, planned maintenance. Leadership did not investigate why these critical components were failing, and during the facility shutdown, leadership postponed or eliminated some of these critical equipment tests and inspections to get the unit started back up.

Relevant RBPS Elements

- Asset Integrity and Reliability
- Hazards Identification and Risk Analysis
- Incident Investigation

LL4) The following points, noted in part from the perspective of those directly impacted by this incident, are summarized in Table 5.3 [8, p. 67].

Relevant RBPS Pillars

Due to the number of this incident's published investigations, this analysis included a brief discussion on the weaknesses to the relevant RBPS Pillars, as follows:

1) Pillar I – Commit to Process Safety
- Process Safety Culture
- Compliance with Standards
- Process Safety Competency
- Workforce Involvement

At the time of the incident, the weak commitment to process safety at the facility was attributed to 35% of the overall CCPS RBPS pillar weaknesses (see discussion in Section 10.3).

2) Pillar II – Understand Hazards and Risk
- Process Knowledge Management
- Hazards Identification and Risk Analysis

At the time of the incident, the weak understanding of the hazards and risks pillar at the facility was attributed to a little more than 10% of the overall CCPS RBPS pillar weaknesses (Section 10.3).

3) Pillar III – Manage Risk
- Operating Procedures
- Safe Work Practices
- Training and Performance Assurance
- Management of Change
- Operational Readiness
- Conduct of Operations
- Emergency Response

At the time of the incident, the weak manage risk pillar at the facility was attributed to almost 50% of the overall CCPS RBPS pillar weaknesses (Section 10.3).

4) Pillar IV – Learn from Experience
  • Incident Investigation
  At the time of the incident, the incident investigation element for both near miss (weak signal) incidents and inadequate investigations to the deeper underlying causes was attributed to 3% of the CCPS RBPS pillar weaknesses (Section 10.3).

As will be discussed more in Chapter 10, the weaknesses in Pillars II, III, and IV can be attributed to overall weaknesses in Pillar I, the specific facility's commitment to process safety.

C5.6.3-2 – Chemical Runaway Reaction, Pressure Vessel Explosion and Fire [50] [51]

Incident Year: 2008

Cause of the process shut-down: Extended outage scheduled to install a new process control system and a new pressure vessel
Cause of incident occurring during the start-up: Impaired human performance due to increased complexities of the new operating computer system and inadequate communication of processing conditions between shifts during restart

Incident impact: Explosion and intense fire that burned more than four hours, causing two fatalities, eight injuries, evacuation of more than 40,000 nearby residents to shelter-in-place locations, and closure of a nearby highway due to smoke disrupting the traffic

Risk management system weaknesses:
LL1) Operators did not receive adequate training on the new computer system combined with operators arriving for the next shift not receiving adequate handover communications. The new shift did not understand that a slow temperature rise was due to the exothermic polymerization reaction, and that once they noticed a

rapid temperature rise, the uncontrolled runaway exothermic chemical reaction was already occurring

Relevant RBPS Elements

- Process Knowledge Management
- Training and Performance Assurance

LL2) The Standard Operating Process (SOP) did not require regular lab sampling of flash bottoms in the residue treater as a safeguard to identify if the concentration was too high [this had been identified in the Process Hazards Analysis (PHA)]. However, the SOP for the start-up included sampling the liquid remaining in residue treater from the prior run before restarting the unit and then before adding new feed. The new shift did not know that the lab had reported a too high concentration in the residue treater before restarting the residue treater.

Relevant RBPS Elements

- Process Knowledge Management
- Hazard Identification and Risk Analysis
- Operating Procedures

(Note: Other Lessons Learned included addressing an ineffective Operational Readiness Review (i.e., the Pre-Startup Safety Review) and emergency response protocols (i.e., Emergency Management) [50])

Table 5.3 Points to remember when starting up after a facility shutdown time

No.	"Points to Remember" (IChemE BP Loss Prevention Series)	Pillar I – Commit to Process Safety (23) [35%]	1 Process Safety Culture (14)	2 Compliance with Standards (2)	3 Process Safety Competency (2)	4 Workforce Involvement (5)	5 Stakeholder Outreach (0)	Pillar II – Understand Haz. and Risks (8) [12%]	6 Process Knowledge Management (3)	7 Hazard Identification and Risk Analysis (5)	Pillar III – Manage Risk (32) [49%]	8 Operating Procedures (6)	9 Safe Work Practices (1)	10 Asset Integrity and Reliability (2)	11 Contractor Management (0)	12 Training & Perform. Assurance (7)	13 Management of Change (2)	14 Operational Readiness (4)	15 Conduct of Operations (9)	16 Emergency Management (1)	Pillar IV – Learn from Experience (2) [3%]	17 Incident Investigation (2)	18 Measurement and Metrics (0)	19 Auditing (0)	20 Management Review and Contain, Improv. (0)
	BP Texas City Incident – CCPS RBPS Sum	23	14	2	2	5	0	8	3	5	32	6	1	2	0	7	2	4	9	1	2	2	0	0	0
1	Start-up/shutdowns are rare, so refresher training may be needed.	2	1		1			1		1	2					1			1		0				
2	Make sure all critical instrumentation/ equipment are functional and that certifications are current.	2	1	1				2	1	1	1			1							0				
3	Make sure that all work permits have been closed and the equipment is approved for use.	2	1	1				0			3		1					1	1		0				
4	Talk through the procedures as a team before each shift.	2	1			1		0			2	1							1		0				
5	Identify hazards and safeguards listed in the procedure.	2	1			1		2	1	1	2	1	1								0				
6	Follow written procedures/checklists step by step.	2	1		1			0			3	1				1			1		0				
7	Make sure all valves/blinds/locks are in the proper position.	1	1					0			3	1						1	1		0				

Table 5.3 Points to remember when starting up after a facility shutdown time (Continued)

"Points to Remember" (IChemE BP Loss Prevention Series)	Pillar I – Commit to Process Safety	1 Process Safety Culture	2 Compliance with Standards	3 Process Safety Competency	4 Workforce Involvement	5 Stakeholder Outreach	Pillar II – Understand Haz. and Risks	6 Process Knowledge Management	7 Hazard Identification and Risk Analysis	Pillar III – Manage Risk	8 Operating Procedures	9 Safe Work Practices	10 Asset Integrity and Reliability	11 Contractor Management	12 Training & Perform. Assurance	13 Management of Change	14 Operational Readiness	15 Conduct of Operations	16 Emergency Management	Pillar IV – Learn from Experience	17 Incident Investigation	18 Measurement and Metrics	19 Auditing	20 Management Review and Contain. Improv.
CCPS Risk Based Process Safety Element	23	1	2	3	4	5	8	6	7	32	8	9	10	11	12	13	14	15	16	2	17	18	19	20
BP Texas City Incident - CCPS RBPS Sum	23	14	2	2	5	0	8	3	5	32	6	1	2	0	7	2	4	9	1	2	2	0	0	0
			35%					12%							49%							3%		
8 Never go to the next step of the procedure before all conditions from previous steps aren't fully completed with satisfactory results.	2	1	1				0			3	1							1		0				
9 Check equipment setup thoroughly and monitor initial conditions in the field, maintain excellent communication with control room.	2	1		1			0			3							1	1		0				
10 Use Management of Change reviews before modifying any start-up procedures.	1	1					1		1	3			1			1		1		0				
11 Report any deviations/anomaly. Keep an open eye for "weak signals."	1	1					0			1								1		1	1			
12 Do not be satisfied by short term fixes ("symptom treatment") but identify and treat the root causes of incidents.	1	1					0			2	1							1		1	1			
13 Ask questions and get help with operations which you are not familiar with.	2	1		1			0			2					1				1	0				
14 If unsure, or if multiple parameters seem to deviate from the usual ones indicated in the procedure, shut-down the process and alert potentially exposed personnel.	1	1					2	1	1	2		1						1		0				

5.7 How the RBPS elements apply

All of the Risk Based Process Safety elements (RBPS) apply when setting up a process safety and risk management program to manage effectively the process safety risks during the shut-downs for facility shutdowns and the start-ups afterward (the Type 5 and Type 6 transient operating modes listed in Table 1.1). Additional guidance for applying and auditing these RBPS elements for an effective overall process safety and risk management program is provided in other resources [14] [31].

Part II
Abnormal and Emergency Operations

6 Recovery

6.1 Introduction

This chapter defines the term "recovery" and how the operations team recovers process control during an abnormal operation (Section 6.2), describes an approach used to help anticipate the process deviations that can result in abnormal operations (Section 6.3), and how abnormal operations can be effectively managed for successful recovery efforts (Section 6.4). The discussion continues with a recovery-related incident and its lessons learned (Section 6.5). This chapter concludes with a brief overview of how an RBPS program and its elements can be used to effectively manage and help organizations improve their responses to abnormal situations (Section 6.6).

6.2 Recovering from an abnormal operation

"Recovery" occurs during abnormal operations when the operating deviations remain within control and do not require shutting the process down. The recovery mode is defined as "the time when operations can safely respond to a process upset and return it to normal operations, recovering the process to its normal operating conditions" (Table 2.1). The flow chart used to help illustrate the operation's team responses to abnormal operations was introduced in Figure 1.3. Although successful recovery efforts do not have a "transient operating mode" associated with them, if the recovery is not successful, then the operations team has to select from two shut-down possibilities depending on how much control they have when shutting the process down safely (Figure 1.3). Sometimes the process can be placed into a standby mode while the process upset is being diagnosed. The standby mode is a temporary, stable, and safe time when the equipment is not at its normal operating conditions and still within its safe operating limits, but it is not completely shut-down, either. Discussions on the unscheduled and emergency shut-downs used when the operations team is not successful in its recovery efforts are described in more detail in Chapter 7 and Chapter 8.

6.3 Anticipating abnormal operations

Process safety-related risks can be anticipated by and evaluated with a Hazards and Operability (HAZOP) study: a methodical, structured, and guideword-based analysis of the process [15]. The structured HAZOP method, used as a part of a Process Hazard Analysis (PHA) study, helps the team anticipate and evaluate the hazards and risks associated with process deviations. An example of a few of the potential process condition deviations for a continuous process is shown in Table 6.1. The guidewords are used to help establish the design intent and to stimulate the discussions by identifying process deviations which could potentially lead to a loss event. As shown in Figure 6.1, process deviations can result in a continuum of responses, from the successful recovery efforts described in this chapter, to an unscheduled shut-down (described in Chapter 7), or—by far the least desired option—to an emergency shut-down (Chapter 8). For example, pressures higher than the expected operating pressure would fall in the "High Pressure" deviation noted in Table 6.1. Well-conducted hazards reviews have helped reduce the percentage of incidents involving incorrect design or use of materials of construction and straight mechanical failures from 69% in the 1950's to 42% in the 1980's [3, p. 4].

Table 6.1 Example guidewords used in a HAZOP study for a continuous process.

Guideword	Flow	Temperature	Pressure	Level (Interface)
More	High Flow	High Temperature	High Pressure	High Level
Less	Low Flow	Low Temperature	Low Pressure	Low Level
None	No Flow	Cryogenic	Vacuum Pressure	No Level
Other Than	Reverse			
	Misdirected Flow			

If the deviation applies to the equipment under review, the multi-disciplined PHA Team will pose and discuss the potential deviation and determine which deviations have process safety risks and need protection layers to help prevent an incident. Thus, abnormal operations that have process safety-related risks during the transient operating mode can be anticipated, planned for, and effectively managed. This PHA Team-oriented approach, including the Hazards Identification and Risk Analysis (HIRA) approach, is described in more detail in other publications [4] [14] [52] [53] [54]. A more comprehensive guideword matrix is presented for both continuous and batch operations in the the Appendixnd another publication [21, p. 194].

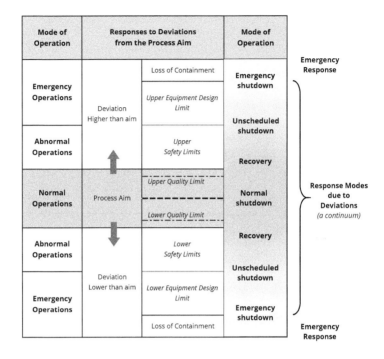

Figure 6.1 Types of responses to process aim deviations.

Although thorough operating-related guidewords provide a comprehensive operational hazards review, incidents have occurred during the transition times due to procedural violations, especially when non-routine tasks are performed or people should respond quickly to an abnormal operation. Thus, guidance has been provided using the HAZOP methodology successfully to address the procedural issues that may occur during these transition times [4]. In particular, the common element leading to incidents in these transition times was "for increased human interaction with the process...[with] the operator and procedural controls...[being] the key layer of protection for preventing an incident. [4, p. 2]" A HAZOP is designed to ensure that sufficient protection layers are in place to reduce the risks, whether they are preventive or mitigative engineering or administrative controls. It is the combination of the effectively designed, implemented, and sustained controls that helps reduce the process safety risks.

Many facilities have started to anticipate security threats and prevent or mitigate the consequences once the facility perimeter has been breached. In particular, geopolitical unrest and wars in countries around the world have made terrorist targets out of refineries, natural gas facilities, and other manufacturers using toxic, flammable, or explosive materials. Cyberattacks have also been successful in reducing productivity as well as jeopardizing the process safety of processes handling hazardous materials and energies. For this reason, there are many defensive tactics that can be implemented to help reduce the likelihood of an attack, and in the event the facility is targeted, designing the emergency response capability to reduce the consequences of the release is key. Additional details are provided in other publications [55] [56, pp. Chapter 35, pp. 2-8] [57] [58].

6.4 Managing abnormal operations

Abnormal operations are defined as "the operating mode that occurs during normal operations when there is a process upset and the process conditions deviate from the normal operating conditions" (Table 2.2). These upsets to the normal processing conditions may result in deviations that can range from relatively small deviations to large deviations that may become too difficult for the recovery efforts to manage. Thus, a "normal" operations ventures into the "abnormal" operations territory, and if the process upset begins to or exceeds the safe operating limits, "emergency" operations are implemented (Figure 6.2). This section describes how abnormal operations can be effectively managed for successful recovery efforts, beginning with a description of the abnormal situation (Section 6.4.1).

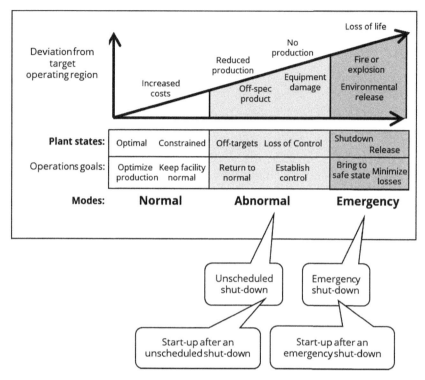

Figure 6.2 Transient operating modes associated with abnormal and emergency operations.

(Adapted from [15])

This section continues with how abnormal situations can be addressed (Section 6.4.2) and concludes with a description of how the operations group can respond to expected deviations with a successful recovery (Section 6.4.3).

6.4.1   Recovering from an abnormal situation

Recovery actions for abnormal situations in both continuous and batch operations are the operating group's response efforts to keep the process under control—to recover from the abnormal situation. During the normal operation, the process is operating within its safe limits for both steady-state continuous and unsteady-state batch processes, as was illustrated in Figure 3.2 and Figure 3.3, respectively. An abnormal situation is defined as "a disturbance in an industrial process with which the Basic Process Control System (BPCS) of the process cannot cope" [34]. The recovery efforts to control successfully the process safety risks depend on the integrity and reliability of the engineering controls and on operational discipline from those responding through the administrative controls [21].

Since relatively "small" process deviations are expected, they can be successfully responded to, as is illustrated for a continuous process in Figure 6.3's timeline. The process deviations can be anticipated and determined beforehand using the hazards and risks analysis approaches, as discussed briefly for a higher pressure deviation in Section 6.3. Then, with the understanding of the deviations with process safety risks that should be managed, the engineering and administrative controls required for normal operations can be identified, designed, implemented, and maintained. These controls help the operations team safely return the process to its standard operating conditions once the deviations are detected. Engineering controls include understanding the dynamic characteristics of the Basic Process Control System (BPCS) [59]. The administrative controls, the procedures for "normal operations" are written to guide the operations team on the standard (expected) operating conditions and often provide minimal troubleshooting protocols to help with the recovery efforts. As was noted earlier in this chapter, when the recovery efforts for an abnormal situation are unsuccessful, then the operations team will shut the process down.

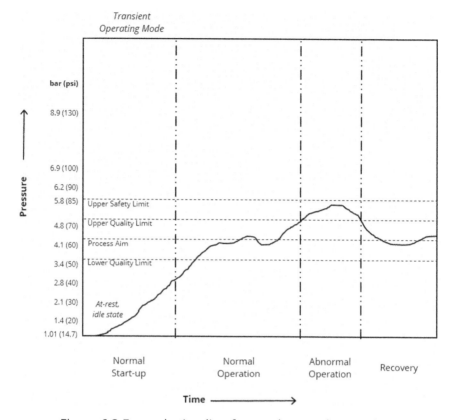

Figure 6.3 Example timeline for an abnormal operation on a continuous process with recovery.

As described earlier, there are different types of engineering and administrative control responses to process deviations; high- and low-process deviations may exceed the safe equipment design limits and lead to loss of containment. Whether it is a continuous or batch process, these engineering and administrative controls help the operation team detect, respond to, and recover from process deviations. Hence, the recovery actions help keep the process under control, effectively managing the risks associated with the hazardous materials and energies by preventing the processing conditions from exceeding the safe equipment design limits. As depicted in Figure 6.3, a successful recovery *does not result* in a processing shut-down, and thus, there are no transient shut-down operating modes associated with this day-to-day experience.

At this point, it is worth noting that a successful recovery will return the process to its normal operation with a normal shut-down at the end of the production run. However, if the recovery is not successful, a continuum of responses to normal operations will occur depending on the extent of the deviation. If the operations group decides to use the normal shut-down procedures, then there is an unscheduled shut-down (described in Chapter 7). If, however, the process should be shut-down immediately, the operations group initiates its emergency shut-down procedures. These emergency operations are discussed in detail in Chapter 8.

6.4.2   Managing abnormal situations

Since abnormal situations may lead to abnormal operations, it is essential that systems be in place to manage them effectively. This section will highlight some of the aspects that need to be written into the troubleshooting efforts, both engineering and administrative, when the operations team responds to deviations and successfully recovers from an abnormal situation. As the deviations are evaluated in a hazard evaluation, the larger the deviation, the larger the response tends to be. The BPCS receives and evaluates field transmitter values of the processing conditions (such as temperature or pressure) and equipment parameters (such as levels and valve open/closed positions). When the BPCS is "in control," the feedback loop maintains the operating conditions within the quality and safe operating limits. When it can no longer effectively control the process, the process is no longer in its normal operations mode, and an abnormal situation is underway.

In addition to the automatic controls monitored by and controlled by the BPCS during normal and abnormal operations, there are alarms that let the operations team know if there is something that requires an administrative action on their part. These administrative controls may be as simple as acknowledging the panel alarm or the response may involve ensuring that someone visually checks or adjusts components or equipment in the field, such as opening or closing a manual valve. The administrative controls include the normal operating procedures, as well as special procedures during the transient operation modes: normal process start-up and shut-down, emergency process shut-down, and start-up after an emergency process shut-down. If the situation is new and there is no specific procedure for the situation, then there should be a situation-response management system in place to address the issues safely that might arise. Typically, the facility's change management system is used to address such temporary operations [14] [21].

Since the normal operating procedures can include general troubleshooting guidance, it is essential that they are well written, up-to-date, and effectively implemented (refer to Section 3.3). However, it is impossible to write a procedure for every potential, unpredictable deviation that might (and sometimes do) occur, whether the deviation occurs during a normal process start-up, normal operations, or when normally shutting the process down. This poses a problem for writing effective troubleshooting guides, as operating procedures for normal operations can be prescriptive, step-by-step instructions for specific tasks when operating the process within its safe operating limits. Procedures can be written for the normal equipment and process start-ups and process shut-downs, the transient operating modes, since the changing, transient operating conditions and their associated hazards and risks are (or should be) understood. For this reason, only general troubleshooting guidance can be expected in the written procedures.

An industrial alliance, the Abnormal Situation Management (ASM) consortium, was formed in 1994 to "create a new paradigm for the operation of complex industrial plants, with solution concepts that improve Operations' ability to prevent and respond to abnormal situations [60]." Their "focus is on management practices that influence the organizational culture, work processes, staff roles and responsibilities, and valued behaviors as they relate to abnormal situations [60]." As is shown in Figure 6.4, with a focus on equipment-related failures, the goal of the ASM Consortium is to detect when equipment failure begins using predictive tests and inspections before its performance degrades. The next detection mode is proactive, where the conditions, such as corrosion rates, can be used to monitor the equipment, usually resulting in moderate equipment damage that can be addressed proactively. If neither of these equipment condition-monitoring steps are in place, operations should react to the degraded equipment with often-costly repairs and emergency maintenance work. If the equipment's condition is not monitored, detected, or even suspected that it's not fit for service, it may catastrophically fail [25] [61]. More discussion on effective approaches to enhance asset integrity programs is provided in Chapter 10.

It is important to note that an "abnormal situation management" program is effective when the abnormal situation can be anticipated and predicted, detected, and responded to effectively. However, abnormal situations that are unanticipated, unpredicted, are usually not detected, and often result in emergencies. Refer to the Appendix for additional discussion on "expecting the unexpected" when addressing unexpected changes which may need to be implemented during start-ups and shut-downs.

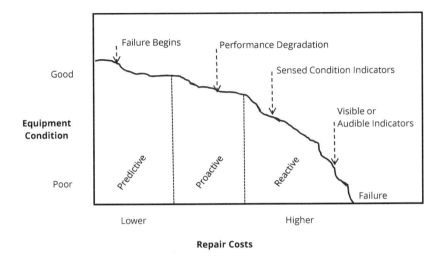

Figure 6.4 Example abnormal situation fault detection model.

6.4.3 Response to expected deviations with successful recovery

The process is designed to make quality products during normal operations of a continuous process; for batch processes, quality product is extracted and separated, as needed, from the batch vessel at the end of the production run. Successful recovery efforts result in continued production, exhibiting the recovery behavior as was illustrated in Figure 6.3 for a continuous process. When everything is working well and as designed, it is a good and safe production day at the facility—in other words, the control of its hazardous materials and energies have been effectively managed.

6.5 Incidents and lessons learned

When there are established, effective, normal recovery procedures, no process safety-related incidents should occur during this time. When abnormal situations repeat, they are likely to be leading indicators of an impending worse scenario. Hence, they should be investigated and addressed before they become uncontrollable [29]. In addition, when incidents occur after personnel think they have successfully recovered the process, special attention to the reasons for the unsuccessful recovery steps are warranted. For example, the following incident occurred when operations thought they had successfully recovered from a short, ten-minute cooling system loss during their normal operations.

C6.5-1 – Perceived recovery that resulted in a runaway reaction [62] [63]

Unfortunately, not all recovery efforts are successful, even if they appear to be. This case describes a batch reactor that had only a ten-minute loss of cooling that resulted in a runaway reaction, reactor overpressurization, and significant explosion.

Incident Year: 1969

Cause of the process upset: Routine, accepted failure of and then recovery from the nitroanaline batch system's heat exchanger within the usual ten minute's recovery time

Cause of incident occurring after perceived recovery: Undetected runaway reaction due to the combination of a nitrochlorobenzene overcharge and an ammonia undercharge; see timeline provided in Figure 6.5 (compare to Figure 3.3, normal isothermal batch operations).

Incident impact: Runaway reaction overpressurized the reactor, which catastrophically failed and exploded. Four people were injured. The facility was destroyed. The explosion was heard almost 16 km (10 miles) away.

Risk management system weaknesses:

LL1) Personnel did not understand the increased risk when they increased the production rate and batch volume. The heat removal design was insufficient to prevent overheating, resulting in the runaway reaction and overpressurization of the reactor.

Relevant RBPS Elements

- Process Knowledge Management
- Hazard Identification and Risk Analysis
- Management of Change
- Operating Procedures

LL2) Personnel and management accepted routine failure of the cooling system since the recovery efforts without a shut-down had always been successful and kept operations running within its normal operating conditions.

Relevant RBPS Elements

- Process Safety Culture
- Asset Integrity and Reliability
- Conduct of Operations

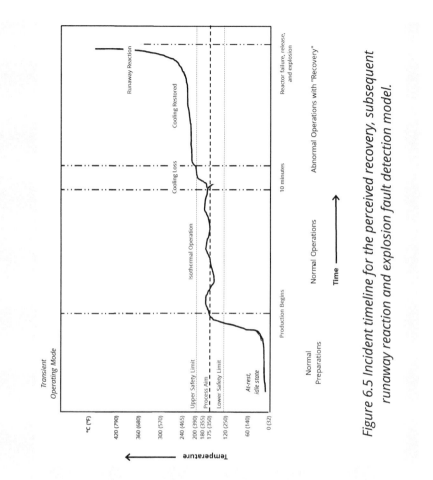

Figure 6.5 Incident timeline for the perceived recovery, subsequent runaway reaction and explosion fault detection model.

6.6 How the RBPS elements apply

All Risk Based Process Safety elements (RBPS) apply when setting up a process safety and risk management program to manage the process safety risks effectively. Effective recovery is a result of an effective process safety program. For successful recovery efforts—the subject of this chapter—it is essential that the hazards be understood, the risks evaluated, and the engineering and administrative controls be identified, designed, implemented, and sustained for the life of the process. Effective process safety and risk management programs are the subject of considerable guidance today, noting that the knowledge of how to identify, design, implement, and sustain the technologies for these programs continues to evolve. Additional guidance for applying and auditing these RBPS elements for an effective overall process safety and risk management program is provided in other resources [40].

7 Unscheduled Shutdowns

7.1 Introduction

This chapter defines the unscheduled shutdown in Section 7.2. How to anticipate and prepare for the unscheduled shutdown is described in Section 7.3. Section 7.4 covers the start-up after the shutdown time is over. Since natural hazards, in particular extreme weather events, have caused significant damage and have resulted in significant incidents, Section 7.5 provides a brief overview of how the shut-downs and start-ups—the transient operating modes—associated with natural disasters have been addressed depending on the location of the facility. Section 7.6 describes incidents and lessons learned from unscheduled shutdowns. In addition, this chapter concludes with a brief overview of how an RBPS program and its elements can be used to manage the risks effectively of the transient operating modes associated with unscheduled shutdowns.

7.2 Unscheduled shutdowns

The unscheduled shutdown is defined as "the time between an unscheduled shut-down and the start-up afterwards. (Table 2.1)." Although the timing of these shutdowns is "unplanned" in the forecasted production schedule, they can be anticipated and properly planned. In other words, there *are plans* in place for an unscheduled shutdown due to the "unplanned" timing of the shut-down.

Unscheduled shutdowns can occur when:

1. The process cannot be successfully recovered from an abnormal situation and the normal shut-down procedures can be used, or
2. There is time to prepare and shut-down the facility for a pending natural hazard (e.g., a hurricane or cyclone).

These shut-downs were illustrated in Figure 6.2, the transient operating modes associated with abnormal operations. If other groups are involved in the unscheduled shutdown, such as maintenance, special permits and handover procedures should be in place beforehand, as needed, before performing the shutdown-related activities.

7.3 Anticipating and preparing for unscheduled shutdowns

There are three options for the operations team to choose from for its process shut-down—transient operating mode Type 7, Table 1.1—before an unscheduled shutdown. These are when the team:

1. Cannot recover the process from an abnormal situation, but has sufficient time to use the normal shut-down procedures;
2. Should prepare the facility for an anticipated or pending natural hazards-activated shut-down (e.g., hurricane or cyclone); or
3. Has to use the shut-down procedures in response to a natural hazard event that has occurred (e.g., an earthquake, lightning strike).

If other groups are involved in the shut-down procedures, special permits and handover procedures should be in place before performing the shutdown-related activities. These shut-down options are briefly discussed next.

7.3.1   Activating a shut-down due to an unsuccessful recovery attempt

The types of shut-downs due abnormal operations are shown in the flow chart in Figure 1.3, with an example timeline of an abnormal operation with a normal shut-down for a continuous process illustrated in Figure 7.1. This differs from the successful recovery to an abnormal situation as was discussed in Chapter 6 (Figure 6.1), as the shut-down occurs in this case from an *unsuccessful* recovery, with the operations team recognizing that they have time and could shut the process down using the normal shut-down procedures. In this case, there is no imminent process safety risk due to the processing conditions approaching the safe operating limits. This could be to an issue the operations group recognize from experience and know that their best option was to shut-down before they lost control of the process, or, more importantly, they recognize that "something's not right" and they need more time to understand the unusual event or events that are occurring better. This is known as keeping a sense of vulnerability (see additional discussion in the Appendix); the operator has the right to shut the process down before something bad happens.

It is important to recognize at this point, as well, that the operations team has other unscheduled shutdown options as is illustrated in Figure 1.3, the abnormal and emergency operations flow chart. These shut-down responses to an emergency shutdown will be discussed in Chapter 8.

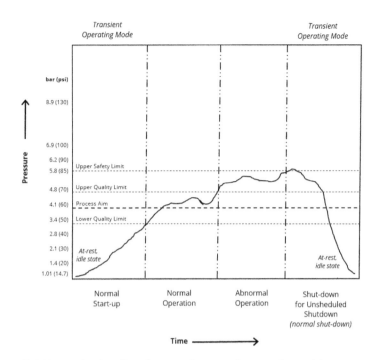

Figure 7.1 Example timeline for an abnormal operation with a normal shut-down (an unscheduled shutdown).

7.3.2   Activating a shut-down due to natural hazards events

Since there are location-specific natural hazards, such as hurricanes or cyclones which can be anticipated, prepared for, prevailed through, and recovered from, the operations team will have additional, specially written plans of the steps to prepare for, assess for damage, and then recover from the event. These additional shut-down activities may be non-routine, and as such, may not be performed very often. A shut-down in preparation for a natural hazards event may take more time than a normal shut-down due to these additional procedures, and as such, may need to be performed at the same time the process is operating before it is shutdown.

7.3.3   Activating a shut-down due to loss of utilities

Although every type of utility loss can be anticipated, some of them, to some extent, cannot be predicted as to *when they will occur.* Anticipated utility losses include scheduled work on the supplier's system. Unanticipated utility loss events could be sudden or rapid (i.e., electricity loss) or slow (i.e., steam loss). Whether the event occurs quickly or not, the facility determines the types of engineering and administrative controls needed to manage the process safety risks effectively during both scheduled and unscheduled shut-down times. The engineering controls, including engineering design, need to be designed and implemented before the event, and the administrative controls will need to be effectively executed at the unpredictable, but anticipated times. The administrative controls are essential when managing the transient operating mode—when shutting the process down, when assessing the damage, and most importantly when restarting the processes after the operational readiness review. A "Loss of Utilities" Checklist is provided in the Appendix to help guide the planning for and safe execution of a facility's loss of utilities so that it can safely manage its risk while the process is being shut-down.

7.4 Start-up after activating an unscheduled shut-down

The start-up after an unscheduled shutdown—transient operating mode Type 8, Table 1.1—is the time when preparing for and resuming operations after an unscheduled shutdown. These start-ups were illustrated in Figure 6.2, the transient operating modes associated with abnormal operations. If other groups were involved in the unscheduled shutdown, such as maintenance, the special permits and handover procedures implemented for the shutdown-related activities should be reviewed and authorized before resuming operations. This includes performing, as needed, equipment Pre-

Startup Safety Reviews (PSSR) and an Operational Readiness Review (ORR) [14]. In addition, if there were non-routine procedures used to shut-down the process beforehand, there likely will be non-routine start-up procedures during this start-up that are designed to reverse them. Note that the start-up after an unscheduled shut-down may take more time than a normal process start-up due to the additional procedures involved.

7.4.1   Start-up after an unsuccessful recovery attempt

Since the operations team recognizes that they have time and could shut the process down using the normal shut-down procedures, the reasons for the process shut-down need to be understood before the process is restarts. Typical investigation methods should be used, such as a root cause analysis, especially if the unsuccessful recovery attempts repeat.

7.4.2   Start-up after a natural hazard event

An example of an unscheduled shut-down due to a natural hazard is the imminent landfall of a hurricane or cyclone on a coastline facility. Once the damage from the hurricane has been assessed, the damaged equipment can be repaired or replaced, as needed, and will need to be recommissioned through an operational readiness review before start-up. These preparation and recovery steps are essential for a safe start-up after a natural hazard event. Natural hazard events are illustrated in Section 7.6.4.

7.4.3   Start-up after a loss of utilities

Both the engineering and administrative controls that were used to shut down the process safely need to be understood after the loss of utilities. After assessing for potential equipment damage, repairing damage equipment, as needed, and then, most importantly, performing an operational readiness review before the equipment is handed over to the operations group, a safe start-up can be executed.

Incidents that occurred after a power outage are illustrated in Section 7.6.3.

Although start-up after a loss of utilities depends on the type of process and the equipment affected, a start-up after the loss of steam should consider some of the following aspects:

- Prevent potential low steam supply pressure issues at restart by staging the equipment start-ups to prevent a high initial steam load on the steam system;
- Ensure that condensate has not collected in the steam lines, draining them if necessary, to prevent "water hammer" and piping damage upon restart; and
- Ensure that steam introduced to steam coils inside storage tanks and vessels is re-introduced once the vessel contains enough liquid to cover the coils upon restart. (*Note:* The liquid introduced to an empty vessel with heated steam coils can flash or vaporize upon contact with the hot coil, potentially causing equipment damage such as an overpressurization due to the expanding vapors. Catastrophic equipment damage can lead to the loss of containment of the vaporized material.)

Although the USDA Guidebook series focuses on small plants that produce food, this particular Guidebook discusses how to prepare for, respond to, and recover from power outage emergencies [64, pp. 15-19].

7.5 Managing unscheduled shutdowns caused by natural hazard events

Major production-stopping, facility-crippling natural hazard events have caused unscheduled shutdowns around the world, with some of them leading to significant process safety-related incidents [36]. A term often used in context of managing the risks of natural hazards is "extreme weather," when the local weather is significantly different from its usual weather pattern [65]. Although every one of the natural hazards can be anticipated to some extent, some of them cannot be predicted as to *when they will occur*. The onset of natural hazards events can be unpredictable (i.e., earthquakes) or tend to be more predictable when they occur (i.e., hurricanes).

Whether the event is predictable or not, the facility's location determines the types of engineering and administrative controls needed to manage the process safety risks during these unscheduled shutdown times. A general framework for the information needed to safely anticipate, design for, and respond to natural disasters is provided in Table 7.1 (Adapted from [66]). The engineering controls, including engineering design, need to be designed and implemented before the event, and the administrative controls will need to be effectively executed when needed, especially during both process shut-downs and start-ups.

7.5.1   Identifying location-specific natural hazards

No matter where the facility is located, there will be natural hazards that can affect the facility's normal day-to-day operation, often resulting in an unscheduled shutdown that affects its production schedule. The common natural hazards that have adversely affected how process safety risks should be managed at onshore facilities include the following:

- Geological—earthquakes, tsunamis, volcanoes, landslides, sinkholes, dam rupture;
- Meteorological—hurricanes or cyclones, tropical storms (wave or storm surges), heavy rainfall, hailstorm, lightning strikes, blizzards, freezing rain/ice storms, heavy snowfall, high winds, sandstorms;
- Hydrological—floods; and
- Climatological—extreme temperatures (freeze, heat), drought, wildfires.

Table 7.1 Framework for anticipating and responding to natural disasters.
(Adapted from [66])

1	Identify hazards
2	Gather data
3	Identify equipment to be assessed
4	Evaluate equipment against design criteria
	4.1 Meet design criteria
	4.2 risk assessment
	1. Risks the natural disaster imposes on equipment
	2. Risk imposed on the personnel, the ecosystem and the community
5	Natural Hazard Emergency Response Plan (NHERP)
	1. Emergency Command Center(s) (ECC)
	2. Authority and responsibilities
	3. Understand hazards
	4. Warning systems
	5. Activation prompts
	6. Staffing assignments (including ride out crew)
	7. Evacuation plans
	8. Interdependency
	9. Utility supplies
	10. Communication systems and protocols
	11. Protection of business-critical equipment and data
	12. Inventory
	13. Access and security
	14. Safety
	15. Supplies and logistics
	16. Equipment checks
	17. NHERP maintenance, quality assessment/quality control
6	Recovery
	1. Stabilize and assess
	2. Secure
	3. Repair
	4. Train
	5. Staff
	6. Locate personnel
	7. Lodging and transportation
	8. Supplies
7	Recommisioning
	1. Operational PSSRs
	2. Safety PSSRs
	3. Training
8	Critiques and Learnings
	1. From drills on the NHERP
	2. From NHERP activation

For example, some location-specific meteorological hazard resources are provided elsewhere [67, pp. Table A-8].

7.5.2 Some approaches for managing the risks of natural hazards

Risk assessments on the facility based on its location should include these hazards identification and risk analysis stages to estimate the risk:

1. Damage the natural hazard could have on the facility,
2. Consequences on the site personnel, community, and environment as a result of the damage from the natural disaster, and
3. Estimate of the likelihood of the consequences.
4. Estimate of the risk

Structured guidance that includes these stages is incorporated in the framework depicted in Table 7.1 when anticipating and responding to natural disasters.

Meteorological, hydrological, and climatological storms that can adversely affect a facility may occur with little or no warning, however there may be some time—ranging from hours to days—before the storm's onset, which can provide adequate shutdown preparation time beforehand. Whether there is much or little time for a safe process shut-down, administrative preparation and recovery procedures should be developed for use before they are needed.

The predictable meteorological events with adequate shut-down time include hurricanes, typhoons, or cyclones, tropical storms (wave surges), heavy rainfall, thunderstorms, ice storms, heavy snowfall, high winds, and sandstorms. Lightning-initiated incidents can occur during a thunderstorm, as the strike can ignite flammable atmospheres, such as those in or around storage tanks, and can destroy electrical distribution networks and equipment due to electrical system surges.

An example of a refinery that suffered a major process upset after a severe lightning strike was documented in the UK [68]. Although the lightning strike began a fire in the crude distillation unit, the ensuing plant disturbances and power interruptions affected the rest of the process where the eventual explosion occurred. The explosion was caused by subsequent failures to manage the plant shut-down and upset safely. Major findings which contributed to the damage from the explosion were similar to findings from a major explosion in a US refinery almost a decade later [46]. These findings included: a control system that allowed more liquid into one unit than was leaving it; a control system that was poorly configured and had confusing alarms as operators responded to the upset; and inadequate maintenance on critical instruments used for control.

Hydrological events include floods with potential property damage and operational downtime in flood plains and coastal areas. The impact from river flooding, when the river overflows its banks, and coastal flooding, when coastal water heights exceed normal tide levels, can be devastating.

The risk-based approach includes addressing the phases for anticipating, preparing for, dealing with, and recovering from a natural disaster. These phases include the:

1. Location-specific engineering designs to protect from the natural hazard, including during:
   a. Construction of the facility and its processes
   b. Operation
   c. Maintenance
   d. Process shutdown
   e. Process restart
   and
2. Administrative plans for specific natural hazards, including preplanning for:
   a. Just before the event *(i.e., safe shut-downs),*

b. During the event,
c. Evaluating the damage,
d. Recovering afterward, and
e. Restart *(i.e., safe start-ups)*.

Some activities discussed for evaluating risks of extreme weather events, successful preparation for, perseverance during, and resuming operations safely afterwards have been detailed in other resources [66] [69] [70] [71] [72]. Although the USDA Guidebook series focuses on small plants that produce food, this particular Guidebook discusses how to prepare for, respond to, and recover from hazardous weather emergencies, as well [64, pp. 20-21].

7.6 Incidents and lessons learned

Details of some unscheduled shutdown-related incidents are included in this section. The incident summary is provided in the Appendix.

7.6.1   Incidents during a shut-down activated for an unscheduled shutdown

C7.6.1-1 Batch Vacuum Still placed on "Standby" [20, pp. 287-288]

Incident Year: Not known

Cause of the standby shut-down activated for an unscheduled shutdown: Downstream unit issues; expected duration 2-3 hours; actual duration 5 days

Cause of incident: Normal product processing issues, usually fixed within a few hours, caused the Batch Vacuum Still operators to place the equipment in its usual standby mode. However, the downstream issues were not resolved until five days later. During the standby time, the condenser on the batch still continued to run and the vent valve opened when the vacuum was broken. After the chart recorder ran out of paper, the temperature of the still was not monitored (*Note:* the paper on the recorder was not replaced; this occurred before digital control systems existed). The "closed" heat transfer control valve actually leaked, raising the boiler temperature to the boiling point level of the still's contents.

Incident impact: 180 kg (0.2 tons) of liquid was released through the open vent valve.

Risk management system weaknesses:

LL1) There was no maximum period designated for the standby mode, and no instructions existed for monitoring the batch still system during its standby mode.

Relevant RBPS Elements

- Operating Procedures
- Asset Integrity and Reliability

7.6.2   Incidents occurring during the unscheduled shutdown time

C7.6.2-1 Distillation Column Shutdown and in "Standby" Mode [73]

Incident Year: 2003

Cause of the shut-down activated for an unscheduled shutdown in standby mode: Upstream processing issues five weeks before the explosion.

Cause of incident: Inadequate isolation of the MNT distillation column due to leaky steam valves.

Incident impact: Explosion propelled debris offsite, including one piece weighing six tons, injured three employees, and started several onsite and offsite fires

Risk management system weaknesses:

LL1) Systems were not in place to: 1) adequately evaluate the hazards of the process, including understanding the thermal instability of the material; 2) ensure that sufficient protection layers were in place (including facility siting and design of control room); 3) apply the lessons learned from a similar batch process to its continuous process containing the same material; and 4) incorporate this information in the operating procedures, safe work practices, and training.

Relevant RBPS Elements

- Process Knowledge Management
- Hazard Identification and Risk Analysis
- Operating Procedures
- Safe Work Practices
- Training and Performance Assurance
- Incident Investigation

LL2) Systems were not in place to prevent isolation valve leakage, including corrosion and erosion evaluations.

RBPS Element

- Asset Integrity and Reliability

LL3) Systems were not in place to alert residents and nearby facilities surrounding the facility during an emergency.

RBPS Element

- Emergency Management

7.6.3   Incidents occurring during the start-up after an unscheduled shutdown

C7.6.3-1 (Introduced in Chapter 2) – Millard Refrigerated Systems, Ltd. [18]

Incident Year: 2010

Cause of the shut-down activated for an unscheduled shutdown: Power outage; duration 7 hours

Cause of the start-up incident: Failure of a roof-mounted pipe due to hydraulic shock upon restart after the power outage. Hydraulic shock is "an abnormal transient condition [a rapid deceleration of liquid] that results in a sharp pressure rise with the potential to cause catastrophic failure of piping, valves, and other components, "...often preceded by audible 'hammering' in...piping" [18, p. 2].

As the operations team restarted the ammonia refrigeration system, an operator manually cleared an alarm that interrupted the blast freezer evaporator's defrost cycle, switching the evaporator directly from its defrost mode to its refrigeration mode without bleeding hot gases from the evaporator coil. The control system continued with its normal sequence and subsequently released the hot gases into the downstream piping containing low-temperature liquid. The combination of hot gas and cold liquid created the pressure shocks that ruptured the piping.

Incident impact: 14,600 kg (32,100 lbs) of anhydrous ammonia was released, with the ammonia cloud drifting across a river (Figure 7.2). More than 150 people reported exposure to the released ammonia, with 32 people admitted to the hospital and four being placed in intensive care.

Risk management system weaknesses:

LL1) The control system contained a programming error that permitted the cycle interruption without addressing the potential for the hot gas to be inadvertently released into the cold liquid-filled system. "After an unintended interruption, process upset, or power outage, refrigeration system operators can avoid the need for manual intervention to the defrost cycle sequence by programming the control system to automatically bleed any coil that was in defrost prior to the power outage upon restart [18, p. 10]." Overall, the company had a weak understanding of its refrigeration system controls and of the potential for hydraulic shock. (See additional resource on "water hammer" [20, pp. 171-174, 197].)

Relevant RBPS Elements

- Process Knowledge Management
- Hazard Identification and Risk Analysis
- Operating Procedures
- Training and Performance Assurance

LL2) The emergency responders attempted for four hours to isolate the ruptured piping during the release. The weak engineering controls included an inadequate system isolation design and no emergency shut-down switch for energized pumps, compressors, and valves to prevent material from being added to the rupture piping to limit the amount released.

Figure 7.2 The area impacted by the anhydrous ammonia release incident at start-up.
[18, p. 4]

Relevant RBPS Elements

- Hazard Identification and Risk Analysis
- Emergency Management

C7.6.3-2 Total power failure at an Olefins Plant [74]

Incident Year: Before 1998

Cause of the uncontrolled process shut-down: Total loss of electrical power that forced the plant into an unscheduled shutdown; duration 30 minutes

Incident Impact: Entire Olefins complex lost power, due to 1) the loss of the separate power feed grids entering the facility from independent geographic locations and, 2) the loss of the steam production and the turbo generators in place to help execute a controlled plant shut-down

In addition, the Olefins complex was supported by a number of ancillary utilities, including steam boilers, flares, and oily water sewer oil/water separators.

Direct effects included coked furnace tubes, pressure relief valves that lifted and did not reseat, flashing liquid flow where none should be, compressor and pump damage, water where none should be, plugged cryogenic equipment, superfractionator contamination, distillation column internals damage, smoking flares, and other environmental releases.

Risk management system start-up strengths:

LS1) Performed a utility blow down and check before restarting after the unscheduled shutdown, including steam, instrument air, cooling water, boiler feedwater, flare, hydrogen, and sour gas systems

LS2) Performed an ancillary unit inspection, focusing on a large compressor, furnaces, refrigeration, and the superfractionators first

LS3) Staged operator and craftsmen presence by starting the units up in sequence: fired steam boilers; rolled power of the process gas and refrigeration compressors; ethane furnaces. Then managed C4 storage, heavy ends processing and compositions (Note: water build-up, an issue not recognized during the restart, has been added to the start-up procedures).

RBPS Element Strengths

- Hazards Identification and Risk Analysis
- Operating Procedures
- Asset Integrity

Risk management system start-up weaknesses:

LL1) Inadvertent miss of the boiler feed water line up during start-up, resulting in a Pressure Safety Valve (PSV) relieving water to the oily water sewer separator that also received benzene-containing streams. The separator overfilled, releasing benzene to the atmosphere at levels requiring cordoning off of the separator area due to exceeded exposure limits. The sewer system overfilled, as well, requiring road area cordoning off due to the presence of hydrocarbons and benzene. The control room air conditioners ingested the vapors had to be turned off due to the odors. In addition, the investigation identified little use of the respirators after detection of the hydrocarbons and benzene.

Relevant RBPS Elements

- Operating Procedures
- Training and Performance Assurance
- Emergency Management

<u>C7.6.3-3</u> Power outage at a Ammonium Nitrate Plant [75, p. Case Study Two]

Incident Year: Before 2012

Cause of the unscheduled shutdown: Power outage; duration 17 hours

Cause of the start-up incident: Generation of an explosive gaseous environment in the neutralizer due to the combination of a relatively concentrated ammonium nitrate solution and the corrosion products inside the vessel

Incident impact: Neutralizer exploded during the start-up preparations, causing overpressure damage to plant building walls, flying debris damage to the adjacent neutralizer, concrete beams and girders, and damage to the attached equipment due to the internal overpressures

Risk management system weaknesses:

LL1) The solution in the neutralizer was not drained to safe location when the process was shut-down, with no time limit before the system should be drained, and then inadequate procedures for the process start-up

Relevant RBPS Elements

- Operating Procedures
- Training and Performance Assurance

LL2) Ineffective inspection of equipment, insulation, and process instrumentation, allowing metal corrosion products to accumulated inside the neutralizer

Relevant RBPS Element

- Asset Integrity and Reliability

7.6.4   Incidents caused by natural hazards events

7.6.4.1 Geological event incident

C7.6.4.1-1   Fukushima Daiichi Nuclear Power Plant [76] [77]

Incident Year: 2011

Cause of the nuclear plant shut-down: Tsunami after major earthquake hit the coast of Japan. A 14 to 15 m (46 to 49 ft.) wave created by the earthquake overwhelmed the seawalls surrounding the plant forty minutes after the quake. The flooding caused significant damage, loss of power, loss of control, and then the loss of reactor containment.

Cause of incident occurring after the shut-down: Complete loss of power resulted in shutting down all six units on site. The seawater flooded the cooling system, damaged the electrical equipment, damaged the emergency diesel back-up generator system, and damaged the back-up batteries. The operators could not monitor or control the temperatures of the units. The reactors in each unit overheated in the following days, releasing radioactive material, exposing the surrounding communities and the environment. Explosions occurred at three of the units.

Incident impact: People within 20 km (12.4 mi) near the plant were evacuated or ordered to shelter-in-place over the three days after the tsunami hit. Although no fatalities were attributed directly to the release, there were reports of an increase in thyroid cancers afterwards.

Risk management system weaknesses:

LL1) Leadership had not ensured that their employees had the knowledge, training, and instructions required to manage their process hazards (i.e., the nuclear reactors). During the event, the roles and responsibilities of the various company representatives, regulators, and agencies involved in the emergency were not clear: the emergency preparedness and crisis management had not been established, resulting in confusion and inefficient management of the emergency.

Relevant RBPS Elements

- Process Safety Culture
- Process Safety Competency
- Stakeholder Outreach
- Training and Performance Assurance
- Emergency Management

LL2) Although the plant was designed to withstand significant earthquakes, the company's risk assessments did not adequately address the severe magnitudes that have occurred in the Pacific's "ring of fire" and had underestimated the impact of a tsunami that could result from such an earthquake. The design of the plant left the Fukushima plant vulnerable to the simultaneous critical back-up equipment failure (i.e., the loss of main power supply, the diesel generator, and the battery back-up).

Relevant RBPS Elements

- Compliance with Standards
- Process Knowledge Management
- Hazard Identification and Risk Analysis

Risk Management Strengths:

LS1) Although the Fukushima Daiichi plant suffered the nuclear meltdown, another nuclear power plant, the Onagawa Power Plant [77], located north of the Fukushima plant, also suffered from a tsunami that hit the plant after the same earthquake. However, the design of this plant prevented the common-cause failure that occurred on the critical back-up equipment at the Fukushima power plant. The Onagawa power plant, operated by a separate power company with a different safety culture, was able to perform a safe shut-down with relatively minor damage. In addition, the Onagawa power plant had been built with a more robust back-up energy supply system such that they maintained control of their reactors during the emergency.

Relevant RBPS Elements

- Process Safety Culture
- Process Knowledge Management
- Hazards Identification and Risk Analysis

7.6.4.2 Meteorological event incidents

C7.6.4.2-1 – Hurricane Georges flooding incident [36, p. 19]

Incident Year: 1998

Cause of the facility shut-down: Hurricane Georges flooded a refinery on the shore of the Gulf of Mexico. The Category 2 hurricane storm surge overtopped the dikes built to protect the refinery, leaving the entire facility submerged under more than 1.2 m (4 feet) of salt water, and due to its slow movement, subjected the refinery to 17 hours of high winds and rain.

Incident impact: Salt water damage occurred to approximately 2,100 motors; 1,900 pumps; 8,000 instrument components; 280 turbines; and 200 miscellaneous machinery items; resulting in replacement or extensive rebuilding of the damaged equipment.

Risk management system weaknesses:

LL1) The older areas of the refinery suffered from the flooding, causing significant property damage due to the equipment's layout. Although most of the refinery had suffered significant damage, the newer control buildings and electrical substations sustained little or no damage as they had been built with their ground floors elevated approximately 1.5 m (5 feet) above grade.

Relevant RBPS Elements

- Process Knowledge Management
- Hazard Identification and Risk Analysis

Risk management system strengths:

The incident occurred in 1998. The newer buildings, electrical substations, and equipment located above grade showed how anticipating for an extreme weather event can help reduce the damage associated with the impact of a storm.

C7.6.4.2-2   Arkema Crosby Flooding [78]

Incident Year: 2017

Cause of the facility shut-down: Hurricane Harvey, a Category 4 storm, made landfall and produced "unprecedented amounts of rainfall, causing significant flooding."

Incident impact: The extensive flooding caused by the rainfall exceeded the equipment design elevations, causing loss of power, the back-up power, and the critical refrigeration systems. Eventually, all plant personnel and residents in a 2.4 km (1.5 mile)-radius area surrounding the facility were evacuated before the organic peroxides decomposed and burned. Twenty-one people sought medical attention from exposure to the fumes; more than 200 residents could not return to their homes for a week.

Risk management system weaknesses:

LL1) Although Arkema had a detailed hurricane preparedness plant protecting workers and property before, during, and after a hurricane, none of the Crosby employees anticipated the amount of rain or flooding level. All of the protection layers identified during the Process Hazard Analysis (PHA) failed during the flooding, a common mode of failure which was not recognized or addressed during the study. The Crosby facility personnel at the time of the flooding were unaware of their insurer's floodplain designations, as well, and industrial guidance, at this point, did not specify recommended heights for locating critical equipment in floodplains.

Relevant RBPS Elements:
- Process Knowledge Management
- Hazard Identification and Risk Analysis

LL2) A major highway that bisected the evacuation zone remained open after the evacuation order was given so that emergency responders could move people and resources in nearby areas affected by the flooding. Unfortunately, the decomposition products from the burning trailers affected twenty-one people who traveled through the fumes.

Relevant RBPS Element:

• Emergency Management

7.6.4.3 Hydrological event incidents

The flooding incidents due to hurricanes have been discussed in Section 7.7.4.2. If a facility is located in a floodplain, the same criteria for the facility design applies: locate critical equipment above the flood levels such that high water does not damage the equipment.

7.6.4.4 Climatological event incident

C7.6.4.4-1– Ice Storm [67, p. 89]

Incident Year: Not available

Cause of the process shut-down: Ice storm struck facility two years after it was built

Incident impact: Total loss of power supply to facility due to an unrecognized, common mode of failure resulting in significant business interruption

Risk management system weaknesses:

LL1) The company was aware that the location for their new facility was subject to freezing weather during the winter. However, the local utility could supply the quantity of power required from two independent generating stations. With significant cost savings to the project, and based on the power supply and reliability, the company decided to use the fully purchased power and not construct an on-site power generation facility. The investigation after the total loss of power showed that the power lines from the two independent generating stations to the processing facility were both located above ground and ran in parallel adjacent paths the last 1,500 feet

(460 m). This parallel path ran along the river, suffering from the same icing effects (the common mode of failure).

Relevant RBPS Element:

- Hazard Identification and Risk Analysis

7.7 How the RBPS elements apply

All of the Risk Based Process Safety elements (RBPS) apply when setting up a process safety and risk management program to effectively manage the process safety risks of the shut-downs activated for and the start-ups after an unscheduled shutdown. An effective process safety program is required to anticipate and prepare for these times. It is essential that the shut-down and start-up hazards be understood, their risks be evaluated, and the engineering and administrative controls be identified, designed, implemented, and sustained for the life of the process to manage the risks. Hazards associated primarily with start-ups include pressure, vacuum, and thermal and mechanical shock [8, p. 7]. Effective process safety and risk management programs are the subject of considerable guidance today, noting that the knowledge of how to identify, design, implement, and sustain the technologies for these programs continues to evolve. Additional guidance for applying and auditing these RBPS elements for an effective overall process safety and risk management program is provided in other resources [40].

8 Emergency Shutdowns

8.1 Introduction

This chapter defines shut-downs that occur during emergencies, whether the operations team has decided that the process should be shut-down immediately during abnormal operations or if there has been a catastrophic release of hazardous materials or energies. A brief discussion follows on how to respond to loss of containment incidents of hazardous materials or energies safely (Section 8.3). Guidance and discussions follow on how to anticipate and prepare for these shut-downs (Section 8.4) and on how to start-up safely after the recovery efforts have been completed (Section 8.5). Section 8.6 describes incidents and lessons learned from emergencies, focusing on those pertinent to the activated shut-downs and start-ups afterwards. This chapter concludes in Section 8.7 with a brief overview of how an RBPS program and its elements can be used to manage the risks of the transient operating modes associated with these shut-downs and subsequent start-ups effectively.

8.2 Emergency shutdowns

A shut-down activated for an emergency shutdown is defined as "the time when the operations team has to abruptly shut the process down using emergency engineering controls and/or administrative shut-down procedures" (Table 2.2). These shut-downs can occur when:

- The process cannot be successfully recovered from an abnormal situation, when the operating conditions are at, or may have exceeded, the safe operating limits, and there has been no loss event;

- There has been a loss event requiring a safe emergency response;
- There has been an unpredictable natural hazard event (e.g., an earthquake, lightning strike).

These shut-downs were illustrated in Figure 6.2, the transient operating modes associated with emergency operations. Depending on the facility's emergency response resources and on the extent of the loss event, the Emergency Response Team (ERT) and the Emergency Response Plan (ERP) may have to be activated. Similar to an unscheduled shut-down, these shut-downs may have special procedures, checklists, and decision aids to address other potentially hazardous conditions that may occur during the shut-down. If there is a loss of containment event, there might be additional PPE required during the emergency response when using the special emergency response procedures, activating any Emergency Shutdown Devices (ESD), and activating the ERT.

8.3 Safely responding to an incident

The goal of an emergency response during and after an emergency shutdown is to ensure that everyone is safe, that the injured, if any, are reached and cared for quickly, and that the loss event causing the shut-down is contained as quickly and as safely as possible. All types of emergency responses to shut-downs are based on thoroughly written and implemented emergency response procedures. The preplanning procedures that help ensure everyone's safety during a loss event includes thoroughly written and implemented plans, such as:

- An Emergency Action Plan (EAP) which describes the actions of everyone, including contractors, when an incident is occurring at the facility, and

- An Emergency Response Plan (ERP) that describes the actions for those managing the responses and to those responding to the emergency.

These written plans ensure that:

- Everyone knows how to recognize hazardous emergencies, such as fires, spills or releases, and how to notify others of the emergency.
- Everyone knows the different types of emergency alarms and how to respond to them safely.
- Everyone knows where to go when the alarm sounds, including their safe evacuation routes and shoulder points, if needed.
- Trained personnel know how to change processes and facility to a safe state, including shutting down parts of it or the whole site,
- All emergency responders are knowledgeable of the hazards, have been properly trained and have sufficient and reliable emergency response equipment and PPE required for a safe response.
- Everyone is accounted for during the emergency.
- Regular, realistic, emergency response drills are scheduled and critiqued (both table top and "live" simulations).

Chapter 3, Section 3.3 contains more information on developing procedures. Additional details on planning for incidents and safely responding to them are provided in other sources [21] [40].

8.4 Anticipating and preparing for shut-downs in an emergency

The shut-down for an emergency shutdown—transient operating mode Type 9, Table 1.1—is the operating time that may require non-routine procedures in addition to the normal shut-down procedures, stopping all or parts of the process and all or parts of its production. Typically, the production personnel may have additional specially written plans of the steps and PPE to shut the equipment down, especially during a loss event of hazardous materials and energies.

As noted in Chapter 7, Section 7.2, and, as illustrated in abnormal and emergency operations flow chart (Figure 1.3), the operations team has more shut down-related options due to an unsuccessful recovery effort. Shutdowns activated during emergencies tend to focus on the larger process deviations from the process aim, as was illustrated with the different operations team responses to deviations during abnormal operations (Figure 6.1). When comparing the range of deviations between the successful recovery efforts (Chapter 6) and the unscheduled shut-down responses (Chapter 7), the continuum of responses illustrates that emergency situations typically occur with significant process deviations, such as when the safe operating limits are approached or exceeded, when there has been a loss event of hazardous materials or energies, and after an anticipated but unscheduled natural hazard event which has just occurred (e.g., earthquake). In general, the larger deviations will warrant quicker responses, especially when the safe operating limits are exceeded. In some cases, the quicker engineered emergency shut-down is executed with a Safety Instrumented System (SIS) or an Emergency Shut-down System (ESS) [79] [80]. This is true whether the emergency shut-down is needed for a continuous or batch process [19]. The immediate engineering and administrative actions are taken and place the process in a safer state.

Thus, when the released material causes fatalities due to toxic exposures or to fires and explosions from ignited flammable releases, the proactive emergency responses are too late to prevent harm. Sometimes the fire or explosion has severely damaged or destroyed the very equipment designed to activate and perform the shut-down or help reduce the consequences of the loss event. For example, ignited releases of flammable materials can result in thermal impact to people and equipment. Fires can prevent emergency responders from being able to access the area affected by the fire. Explosions can destroy the emergency response equipment due to damaging overpressures, such as fire monitors and fire water supplies, as well. An example of an explosion severely damaging the emergency response equipment is provided elsewhere [81].

For the purposes of this guideline, there are four general types of abnormal situations that warrant quick, emergency shut-down, usually in this order based loosely on magnitude of the loss event:

1. When the Safety Instrumented System (SIS) or Emergency Shutdown System (ESS) are activated (before a potential loss event);
2. When the SIS or the ESS are activated (after loss event);
3. When there is a loss event (but no activation of the ERP); and
4. When a loss event activates the ERP.

These four types of emergency shut-downs are shown Figure 8.1.

Once the shut-down has been completed, there are two general conditions for the process equipment's end state. Both of these end states are described next: the first is when the emergency shut-down ends at the normal shut-down end state (Section 8.4.1); the second is when the end state is different from the normal shut-down end state (Section 8.4.2).

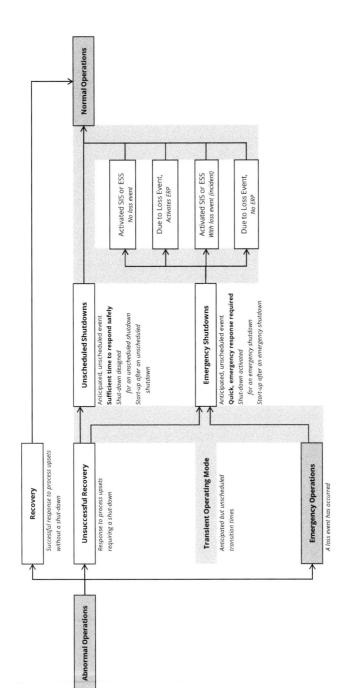

Figure 8.1 Abnormal and emergency operations flow chart with different types of emergency shut-downs

Note for Figure 8.1: Many facilities activate their Emergency Response Team (ERT) when there is a shut-down due to an emergency, especially if there is a loss of containment. This may or may not activate the Emergency Response Plan (ERP) depending on the impact of the loss event.

8.4.1   Emergency shut-downs terminating at the normal end state

The first end state, when the activated shut-down ends at the normal at-rest, idle state, is shown in Figure 8.2. Many facilities activate their Emergency Response Team (ERT) when there is an Emergency Shutdown, especially if there is a loss of containment. This may or may not activate the Emergency Response Plan (ERP) depending on the size and impact of the loss event. The end state is the same as the normal shut-down end state (Figure 3.2). This at-rest, idle state assumes that the activated shut-down for the emergency did not have any additional engineering or administrative controls activated that would affect the condition of the equipment upon its shut down.

8.4.2   Emergency shut-downs terminating at a different end state

The second end state, when an emergency shut-down does not end at the normal at-rest, idle state, is shown in Figure 8.3. If the end state is not safe (as might be in an unanticipated emergency), a thorough assessment of any residual hazards and risks should be addressed before start-up. In addition to potentially activating the Emergency Response Team (ERT) and the Emergency Response Plan (ERP), there may be other engineering and administrative controls that are activated, as well, quickly helping to reduce the impact of the loss event. Since this end state is not the same as the normal shut-down end state (Figure 3.2), personnel restarting the equipment should understand what happened during the activated shut-down, especially if the condition of the equipment was affected during the quick shut down efforts. How to safely start-up after this type of emergency shut-down will be described in more detail in Section 8.5.2.

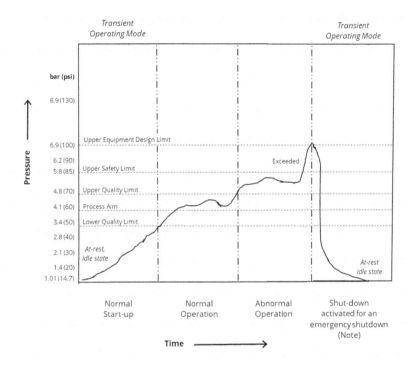

Figure 8.2 Example timeline for an abnormal operation resulting in an emergency shut-down to the normal at-rest, idle state.

Although Emergency Shutdown Device (ESD) designs should address each type of emergency, they may not work if the ESD is impaired or if the ESD has not been properly maintained.. Ineffective asset integrity programs on the ESD and its system – or any mitigative control that is activated after a loss event - may compromise the reliability of the system when it is needed. An impaired ESD may occur, for example, when the loss event damaged important components of the ESD A compromised system will not be able to control or reduce the magnitude of the event. Additional guidance for effective asset integrity programs on emergency shut-down valves and systems, including safety-critical equipment, such as safety controls, alarms, interlocks, and Safety Instrumented Systems (SIS), are provided in other publications [2, pp. 329-350] [21] [23] [79] [76] [82].

In addition, some emergency shut-downs may cause damage to the equipment it is protecting due the consequences of not shutting the equipment down result in the normal manner. There may be physical stress placed on the equipment during the shut-down due to drastic pressure or temperature changes used to bring the process down quickly to a safe end state level that may be nearer to the safe operating limit, but not at the normal shut-down's end state (Figure 8.3). When an ESD is used, any potential, cascading domino effects should be addressed both upstream and downstream of the process or equipment being shut down, as well. Thus, it is essential that everyone responding to the activated ESD understand the condition the equipment, assess for potential damage (which could be internal and, thus, not visible from the outside), and ensure that the equipment is repaired, if needed, before resuming start-up.

8.4.3   Equipment damage due to the emergency shut-down

In some situations, as noted for Emergency Shutdown Device (ESD), the emergency shut-down procedure may cause damage to the equipment it is protecting due the consequences of not shutting the equipment down in the normal manner. For example, steam snuffing systems for a fired burner, firebox, or furnace, when activated, can and do cause internal damage. The physical damage on the firebox internals during the shut-down should be inspected and fixed, as needed, before restart. Thus, it is essential that everyone responding to the emergency shut-down understands the final condition of the equipment, assesses for potential damage, and ensures that the equipment is repaired, if needed, before resuming start-up.

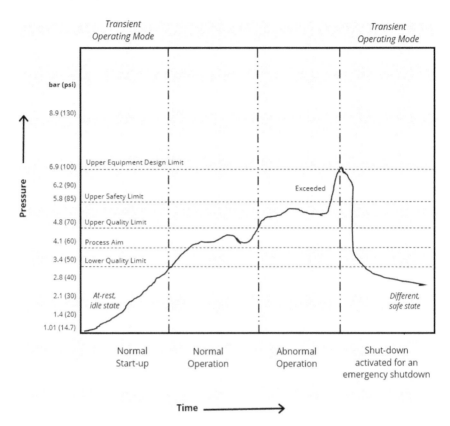

Figure 8.3 Example timeline for an abnormal operation resulting in an emergency shut-down to a different, safe state.

8.5 Start-up after an emergency shutdown

The start-up after an emergency shutdown—transient operating mode Type 10, Table 1.1—is defined as "the time when preparing for and resuming operations after the emergency shutdown period" (Table 2.2). These start-ups were illustrated in Figure 6.2, the transient operating modes associated with emergency operations. If the emergency shutdown time occurred due to an incident with harm to people, the environment, or property, then all special restart and operations-related projects should be designed and implemented before resuming operations. This includes performing, as needed, equipment Pre-Startup Safety Reviews (PSSR) and Operational Readiness Reviews (ORR). The start-up after an emergency includes the operating time that may require non-routine procedures, in addition to the normal start-up procedures, before resuming all or parts of the process and all or parts of its production. A start-up after an activated shut-down may take more time than a normal start-up due to additional engineering or administrative controls which were activated during the emergency. The condition of the equipment for a safe start-up depends on its end state after the emergency shut-down: whether the process was shut down to the normal end state (Figure 8.2) or to different end state (Figure 8.3). The start-up will depend on the facility passing all of its equipment's fit-for-service specifications in its Operational Readiness Review (ORR), especially if damage from the incident was significant. These different end state-related start-ups are discussed next.

8.5.1   After the shut-down to the normal end state

If the normal shut-down procedure is used for the "emergency shut-down" (recognizing that some companies use the same, safe shut-down procedure whether it's an emergency or not), then the equipment should be in its "normal" end state and, thus, should be in a condition acceptable for its safe restart. Sometimes, for example, one piece of equipment may catastrophically fail, and cause the unscheduled, emergency shut-down of the rest of the process unit, other process units, or even the entire facility. The unaffected and undamaged equipment, depending on the extent of the loss event's impact, may not need any special inspections or repairs before start-up since they were shut-down to the normal end state using the normal shut-down procedures.

8.5.2   After the shut-down to different end state

However, if the emergency shut-down procedure results in an abnormal equipment end state, such as rapid discharge of a reactor's contents to a temporary tank or other safe location, then the discharged contents should be safely handled and removed from the affected process equipment or secondary containment before restart. If the equipment's condition is not evaluated before restart, and as noted earlier, it is essential that everyone understands the final condition the equipment to ensure that it is prepared and ready before resuming start-up.

If a part of the facility goes into a circulation or a standby mode while other parts are shut-down, the hybrid operational state can be defined as a different end state, as well. Lessons Learned from incidents that occurred when trouble-shooting other possibly shut-down parts of the process with other equipment placed into a standby mode were illustrated in Chapter 7 (e.g., C7.6.1-1 and C7.6.2-1).

8.5.3   Addressing damaged equipment and processes after a significant incident

If emergency shut-down procedure or Emergency Shutdown Device (ESD) causes damage to the equipment it is protecting due the consequences of not shutting the equipment down in the normal manner, there should be a procedure for addressing the equipment damage before restart. Thorough tests and inspections, especially to check for potential (or known) internal damage, should be performed before the equipment is restarted. Thus, it is essential that everyone understands the final condition the equipment, that the equipment is assessed for potential damage, and that the equipment is repaired, if needed, before resuming start-up.

8.5.4   Rebuilding equipment and processes after a major incident

After a major fire, explosion, flood or hurricane, there will be significant property damage that will need to assessed, with equipment repairs or replacement required before the facility can resume operations. If the affected processes will be rebuilt, the post-incident projects may have considerable business pressure to re-establish production due to commitments to supply products to customers safely and quickly. Companies may decide to duplicate the original design specifications for the rebuild. However, if whatever materials are available on short delivery are procured, even if the specifications are not identical to the original process equipment, the facility needs to have an effective management of change system to ensure that the process safety risks, as well as the demolition and construction risks, associated with the changes can be effectively managed. An additional issue that may need to be addressed includes designing and constructing the replaced equipment to meet the current design standards since older equipment may have been build years ago. Additional discussion for effectively managing engineering projects, especially when significant equipment is being rebuilt, were described briefly in Chapter 4.

8.6 Incidents and lessons learned

Details of some emergency shutdown-related incidents are included in this section. The incident summary is provided in the Appendix.

8.6.1   Incidents during an emergency shut-down

General note: Although the emergency shut-down resulted in an incident, the severity of the incident was reduced due to the mitigative engineering or administrative controls usually activated at the time of the incident.

8.6.2   Incidents occurring during the emergency shutdown time

<u>C8.6.2-1</u> – DPC Enterprises, L.P. [83]

Incident Year:2002

Cause of incident occurring during the emergency shut-down: Upon activation, the emergency shutdown system (ESS) ball valve did not work and did not stop the chlorine release.

Incident impact: Failure of a chlorine railcar unloading hose resulted in release of 21,800 kg (24 tons) before emergency responders could stop release. 66 people sought medical evaluations; three were hospitalized. Trees and other vegetation surrounding the unloading station were damaged.

Risk management system weaknesses:

LL1) At the time of the incident, DPC did not have an adequate Inspection, Testing, and Preventive Maintenance (ITPM) program to ensure asset integrity and reliability. In particular: 1) the transfer hoses did not meet design specifications (there was no "positive materials identification" program); 2) the Emergency Shutdown System (ESS) ball valve did not work when needed due to severe build up (it had not been tested).

Relevant RBPS Element:

- Asset Integrity and Reliability

LL2) At the time of the incident, DPC did not have a clear emergency response plan, did not provide adequate accessibility to its emergency response equipment, did not perform emergency response drills, and had not involved the local emergency response planning committee.

Relevant RBPS Element:

- Emergency Management

8.6.3   Incidents occurring during start-up after the emergency shutdown

General note: In some of the cases, the emergency shut-down resulted equipment damage that went undetected and upon start-up manifested as a loss of containment and control of hazardous materials or energies.

C8.6.3-1 Honeywell Hydrogen Fluoride (HF) Release [84, pp. 73-84]

Incident Year: 2003

Cause of incident occurring during the emergency shut-down: Hydrogen Fluoride (HF) vaporizer rapidly shut-down using emergency procedures when an incident occurred in another unit. The vaporizer was left in an abnormal idle state, containing HF liquid afterwards.

Incident impact: Several weeks later, operations attempted to remove residual HF liquid, releasing HF that injured one employee and exposed an operator.

Risk management system weaknesses:

LL1) At the time, the new system for removing HF liquid in vaporizer did not work a different procedure was used. The HAZOP study had not addressed the deviations from the normal shut-down procedure and the commissioning of the new HF draining system had not been completed. Staff involved in the modified draining procedure did not wear the proper PPE when performing the activity.

Relevant RBPS Elements:

- Hazard Identification and Risk Analysis
- Operating Procedures
- Safe Work Practices

8.7 How the RBPS elements apply

All of the Risk Based Process Safety elements (RBPS) apply when setting up a process safety and risk management program to manage the process safety risks effectively. Effective anticipating for and activating shut-downs during emergencies is a result of an effective process safety program. For safe shut-downs at this time—the subject of this chapter —it is essential that the hazards be understood, the risks evaluated, and the engineering and administrative controls be identified, designed, implemented, and sustained for the life of the process. Effective process safety and risk management programs are the subject of considerable guidance today, noting that the knowledge of how to identify, design, implement, and sustain the technologies for these emergency response programs continues to evolve. Additional guidance for applying and auditing these RBPS elements for an effective overall process safety and risk management program is provided other resources [40].

Part III
Other Considerations

9 Other Transition Time Considerations

9.1 Introduction

This chapter introduces the types of projects that are associated with other transition times, when the processes are not in normal operations, providing an overview of the equipment and process unit life cycle (Section 9.2) and the transition times associated with these life cycle-related projects. The commissioning and initial start-up projects, the first transition time discussed in this chapter, involves new equipment, new process units, or greenfield facilities (Section 9.2.1). The second transition time involves the end-of-life projects (Section 9.2.2). Effective handovers between groups, as noted earlier in this guideline, are essential for effectively managing the process safety risks during these transition times, as these projects often have specialized contractors who have specialized technological or decommissioning expertise (Section 9.2.3).

Special commissioning and initial start-up considerations are discussed in Section 9.3, followed by a review of incidents and lessons learned during the transition time between construction and start-up (Section 9.4). Section 9.5 provides guidance on specific end-of-life shut-down considerations. The two decommissioning-related transition times discussed next are the temporary—mothballing—or the permanent, decommissioning shut-down times. Mothballed processes and equipment are temporarily shutdown for an unknown period of time, requiring some form of preservation during the shutdown period (Refer to Section 5.3.4). Mothballing considerations are covered in Section 9.6, followed by a review of incidents and lessons learned which occurred during these transition times in Section 9.7. Specific decommissioning considerations are discussed in Section 9.8, with corresponding incidents and lessons learned provided in Section 9.9. This chapter concludes with a discussion on how the RBPS elements apply to the start-ups and shut-downs associated with these specific transition times in life cycle-related projects.

9.2 A Life Cycle overview

As was noted earlier, there are transition times in a project's life cycle that have had significant incidents due to lack of proper project management during the handovers (Chapter 4). The two transition times discussed in this chapter are the initial start-up time, when the equipment or processes are being started up for first time, and when the shut-down is being performed for the last time, when the equipment or process is to be mothballed or permanently removed. As was noted earlier, a well-conducted hazards review, such as a HAZOP, helped reduce the percentage of incidents involving incorrect design and straight mechanical failures [2, p. 4]. The project-related transition times, the initial start-ups and final shut-downs, are depicted in Figure 9.1 on the illustration of the project life cycle (Figure 4.2). Thus, the initial start-up stage, depicted as "Startup" project stage #6 in Figure 9.1, is when the chemicals and energies are being introduced for the first time. The shut-down activities associated with the "End of Life" projects are depicted in project stages #10 and #11. These project-related transition times during the project's life cycle—the initial start-up and final shut-down—will be discussed in context of the equipment and process life cycle next.

An equipment and process life cycle can be represented with the eight distinct stages as illustrated in Figure 9.2: 1) design; 2) fabricate; 3) install; 4) commission; 5) operate; 6) maintain; 7) change; and 8) decommission. (Sometimes the fabricate and install stages are combined and referred to as the construct stage. See definitions summarized in Table 9.1.).

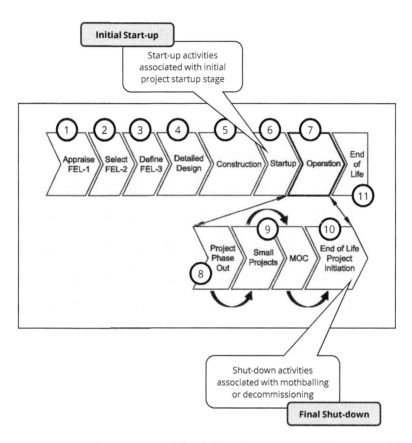

Figure 9.1 Initial start-up and final shut-down stages in the project life cycle.
(Adapted from [31, p. Figure 2.2])

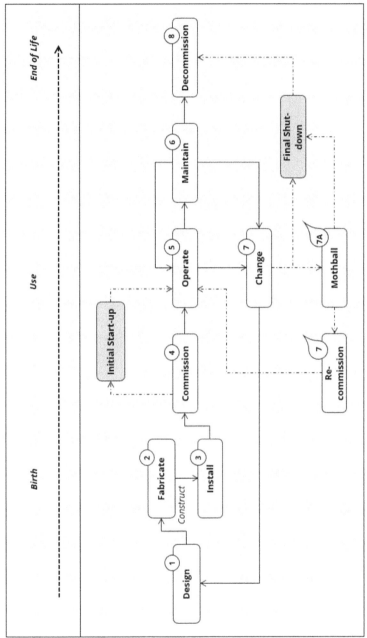

Figure 9.2 Example project life cycle stages in context of the equipment or process life cycle stages.

(Adapted from [21, p. Figure 5.1]*; and* [85, p. Figure 6]*).*

Table 9.1 Definitions of the Process or Equipment Life Cycle Stages.

	Life Cycle Stage		Definitions of the Process or Equipment Life Cycle Stages
1	Design		When the engineering design concepts, the process design parameters, and the equipment design specifications are established and the process knowledge and the process design basis are documented.
2	Fabricate	Construction	When the equipment is fabricated per the fabrication design specifications *(Note: fabrication may occur at the equipment's location in the process unit)*
3	Install		When the assembled equipment is installed at its designated location in the process unit per the installation specifications.
4	Commission		When the installed equipment is approved for safe operations. *Commissioning steps include:* *1) verifying that the equipment and process unit meet their performance specifications:* *1.1) testing the equipment, the control systems, the protection layers, and the utilities,* *1.2) training all operations and maintenance personnel on their tasks and procedures, and* *2) safely introducing the chemicals to the equipment and process units.*
	Initial start-up		The transient operating time when the process chemicals are introduced to the equipment and process units for the first time after the *new, unused or modified* equipment has been fabricated and installed.
5	Operate		When the equipment and the process units are safely operated
6	Maintain		When the equipment is safely maintained using the established, scheduled Inspection, Testing, and Preventive Maintenance (ITPM) program.
7	Change		When proposed changes to equipment design, process design, engineering controls, or administrative (procedural) controls are reviewed, approved, and prepared for commissioning.
8	Decommission		When the equipment's or process unit's useful life is over and the decision has been made to remove the equipment or process unit from normal operations (its end-of-life stage).
	Shut-down for decommissioning		The transient operating time when the process chemicals are removed from the equipment and process units and the equipment are prepared for their end-of-life project stages.
	Mothball		*A decommissioning stage:* When the equipment or process unit may be potentially re-commissioned at a later date.
	Dismantle		*A decommissioning stage:* When the equipment or process unit is dismantled and individual components of the equipment or individual equipment from the process unit may be re-used.
	Demolish		*A decommissioning stage:* When the equipment or process unit is essentially dismantled and for scrap and or material recycling.

The initial start-up of equipment (the smaller projects) and the initial start-up process units or facilities (the larger projects) can be captured in between commissioning (process life cycle stage 4) and operations (stage 5). If the equipment or process unit changes (stage 7) are due to the ultimate end of their in service life (the "final" shut-down), they are dismantled and decommissioned (stage 8). However, when it is uncertain how long the equipment or process unit will be inactive, not used, and idle, the changes are reflected in the mothballing stage, with the equipment or process units either being re-commissioned and operated again (stage 5), or ultimately dismantled and removed at a later time (stage 8).

If a company has decided that the equipment will be partially dismantled or dismantled-in-place when it has mothballed its equipment or processes, it is essential that the equipment's condition is assessed and then properly addressed before attempting to reuse it. Shown in Figure 9.2, these mothballing "steps" would be captured in stage 7A, with the proper Operational Readiness Review (ORR) performed in stage 7B before re-commissioning the equipment and beginning operations again in stage 5. Incidents can occur if the equipment is not returned to its original design or is not re-designed to meet its new service requirements.

This chapter will briefly cover some of the administrative controls which should be developed, reviewed, authorized, and then safely executed to help manage any additional risks associated with initial start-ups (Section 9.3). Section 9.4 will briefly discuss changes for the shut-downs associated with mothballing or decommissioning the equipment or processes. It is essential that an effective change management program be in place to understand any special hazards associated with these transition times and to manage the risks associated with them effectively.

9.2.1   Planning for new equipment or greenfield projects

Planning the projects for new equipment and for development on undeveloped areas—a "greenfield"—follows the similar disciplined project management approach discussed for smaller projects in Chapter 4 and for larger projects in Chapter 5. However, there are constraints that may need to be addressed before start-up for effectively managing the hazards and risks of projects with new, unused equipment, reused equipment that may have been in a different service previously, and greenfield projects. These include administrative controls that should be in place to ensure that the chemicals are introduced safely the first time. In particular, the initial start-up-related planning constraints, activities and controls, designed and implemented during the project start-up stage depicted in Figure 9.1, will be discussed in Section 9.3. A disciplined project management approach, including robust and effective handover controls, applies to both small projects for new equipment and for large greenfield-related capital projects, which may have relatively short duration (the smaller projects) or which may occur over weeks or months (the larger projects).

9.2.2   Planning for end-of-life projects

The planning for an end-of-life project begins once the decision has been made to remove the equipment or process unit/facility from normal operations. The "final" shut-down steps will depend on whether the existing equipment will be mothballed with temporary provisions or fully decommissioned permanently [31, p. 247]. Mothballing occurs when the equipment or process unit may be potentially re-commissioned later. The mothballing-related planning activities and controls, designed and implemented during the project end-of-life initiation stage 10 and the final decommissioning activities in stage 11, will be introduced in Section 9.5.

Decommissioning includes deconstruction, when the equipment or process unit is dismantled and individual components of the equipment or individual equipment from the process unit may be re-used, and demolition, when the equipment or process unit is dismantled for scrap and or material recycling. When a project calls for equipment to be decommissioned, transported to another site and recommissioned, the challenges brought about are a combination of those noted in this chapter plus the transportation-related issues.

As was noted earlier, if a company has decided that the equipment will be partially dismantled or dismantled-in-place when it has mothballed its equipment or processes, it is essential that the equipment's condition is assessed and then properly addressed before attempting to reuse it. As shown in Figure 9.2, these mothballing "steps" would be captured in stage 7A, with the proper Operational Readiness Review (ORR) performed in stage 7B before re-commissioning the equipment and beginning operations again in stage 5. Incidents can occur if the equipment is not fit for its intended service.

Due to the different hazardous materials being handled, planning for the decommissioning project, including hazard identification and risk management, should be in place to reduce the potential for incidents that may cause injury and environmental damage. This includes establishing a project management team, including consultants and contractors experienced in decommissioning, to manage the stages of the project. Decommissioning and end-of-life of equipment or facility may prompt higher focus on cost savings and subsequently project-related short cuts. These short cuts inevitably raise the risk profile and careful attention should be undertaken to avoid this mind-set.

As was noted earlier, handovers between engineering, operations, maintenance, and the specialized decommissioning contractors should be clearly established and safely executed. Some of the handover topics during the stage gate review team that can be used between the end-of-life project initiation at stage 10, Figure 9.1, and the end-of-life decommissioning stage 11, are noted in Table 9.2.

9.2.3 Contractor considerations

Since companies may be unfamiliar with managing new greenfield projects or, in particular, the less-frequent end-of-life projects, contractor selection and oversight is essential to ensure that qualified and competent contractors are used. Effective contractor management during any project was briefly discussed in Section 4.3. Contractors needed to undertake both initial start-up and end-of-life decommissioning projects will require special skills. Typical contracting staff used in normal operations may not be suitable for special tasks performed during an initial start-up or decommissioning projects due to their unfamiliarity with the process safety hazards and risks. The specialized contractors, if needed, will provide the skills required to manage the initial start-up activities and the end-of-life shut-down activities safely.

9.3 Commissioning and initial start-up considerations

The initial start-up is the time when the process chemicals are introduced to the equipment and process units after new unused equipment has been fabricated and installed [31, p. "Startup" xxxi]. Since significant incidents have occurred during initial start-up activities, the following issues should be considered (Adapted from [31, p. 214]):

- Establish thorough handover communications to the operations team, including ensuring that detailed operating procedures have been developed, reviewed, and authorized before commencing start-up.

Table 9.2 Handover topics between the end-of-life project initiation stage to the end-of-life stage. (Adapted from [31, p. Appendix G]).

Stage 11	Project Life Cycle Stage Gate Review From the End-of-Life Project Initiation Stage to the End of Life Stage
	Handover Topics
11.01	Confirm that project plans for decommissioning are adequate.
11.02	Confirm that the operations group is involved as necessary in preparation for decommissioning activities.
11.03	Confirm that the Hazards Identification and Risk Analysis (HIRA) study(s) is(are) complete and recommendations are being satisfactorily addressed.
11.04	Confirm that appropriate specialist reviews have been carried out and their outcomes are being satisfactorily addressed, including engineering controls and checks are in place.
11.05	Confirm that a process safety management system including a process safety plan has been implemented effectively.
11.06	Confirm that an emergency response plan(s) has been developed and that it addresses relevant process safety risks associated with the decommissioning.
11.07	Confirm that process safety aspects have been adequately considered and are appropriate for the decommissioning.
11.08	Confirm that decommissioning workforce training, competency, and performance assurance arrangements are adequate and have been implemented.
11.09	Confirm that the decommissioning project team has a robust process to manage the interface and handovers with contractor(s).
11.10	Confirm that asset integrity management processes including quality management are sufficient to maintain structural and equipment integrity of equipment not involved in the decommissioning.

- Include, if appropriate, representatives from the technology licensors, suppliers, or vendors during the start-up to help support the start-up activities.
- Ensure regular communications between everyone involved in the start-up.
- Establish a barricaded exclusion zone around the equipment or process unit area to prevent non-essential personnel from inadvertently entering the area.
- Field verify valve alignment, energy isolation, feedstock, utilities, etc. before commencing start-up.

- Progress slowly, step-by-step, through the startup operating procedure, including "hold points" in the procedure before initiating a step with a potentially larger consequence.
- Provide extra operations team members to patrol the areas with the new equipment to detect and help respond to abnormal start-up situations.
- Provide quick emergency response protocols to stop the start-up project when an abnormal start-up situation is detected, including communications to those in adjacent areas not involved in the start-up (e.g., during Simultaneous Operations (SIMOPS)) and establishing control of the emergency.

Incidents during the transition time associated with the initial start-up, discussed in Section 9.4, have occurred due to, at least in part, lack of consideration of these items. Some companies may perform a formal "Hazards of Construction (HAZCON)" review before start-ups, as well, to help ensure that the start-up is incident-free. Additional guidance on inspecting equipment in preparation for machinery start-up, described as Testing and Commissioning (T&C), also discusses using Hazards Analyses (HAZAN) and Hazard and Operability studies (HAZOP) is detailed in another reference [86, p. 7].

9.3.1 Planning for commissioning and initial start-ups

Although it may not be possible to plan projects sufficiently or in enough detail in the beginning (especially for the major projects), it is important that careful and detailed planning is performed to ensure safe and efficient commissioning and start-up of new equipment, process units, and facilities. This planning begins during detailed design stage and is refined throughout the construction stage. In addition, process hazards and risk analyses should be performed to help ensure that the commissioning and start-up activities can be completed safely. These studies consider safe introduction of hazardous materials and potential impact on surrounding process units, especially during Simultaneous Operations (SIMOPS), when the commissioning and start-up activities are being performed near operating equipment and processes.

Guidance of these projects should be coordinated by a commissioning manager who oversees the development of and detailing for a commissioning and initial start-up plan and estimates of the budget needed to execute the plan. A typical plan for a greenfield project should include, but not be limited to, the following (Adapted from [31, p. Table 9.1]):

- The project's scope—the equipment, process unit, or facility being commissioned;
- The commissioning and initial start-up team—the number of personnel, required competencies, roles & responsibilities, etc.;
- The training requirements for the commissioning and initial start-up team (e.g., operators, mechanics, electricians, engineers, and other technicians, including classroom training covering the technical, operational, and maintenance guidance and vendor instructions on the new equipment and processes, etc.);
- The contracts for third party support (e.g., technology vendors, engineering design, etc.);
- The day-to-day commissioning and initial start-up resource requirements (such as testing and verification equipment, radios, PPE, water and food, etc.);
- A detailed, individual task level schedule and prioritized sequence of systems that will be inspected, prepared, have chemicals introduced, and operated. (*Note*: These tasks either (i) verify that equipment or a system functions as intended or is ready to operate or (ii) involve operating the equipment, systems, or parts of systems. In addition, scheduling "hold points" should be established, as certain tasks should be completed before other tasks begin, such as commissioning a flare system before the process units.);

- The specific commissioning steps and the reviewed and approved procedures for equipment and process start-up steps, which include the safe operating limits, the consequences of deviation, etc. (*Note*: These procedures are, at best, thorough drafts that typically need updating based on the situations detected and resolved during the commissioning and start-up execution. Any changes should be reviewed and approved through a change management system.);
- And, once normal operations have been achieved for the continuous or batch process (see definitions in Table 2.1), performing test runs to verify that the performance goals, such as throughput and product quality, have been achieved.

Additional guidance when preparing for the initial start-up of a process unit includes using a checklist to inspect the following [87]: personnel safety, vessels, heat exchangers, columns, reactors, piping, machinery, electrical, and instrumentation. Rotating equipment checks include rotational direction, bearing temperature, and vibration during the "run-in" of the motors before they are coupled to their respective drives. If hydro testing or moist air was used, they may hold residual water that needs to be removed by drying the system, especially if the system needs to be dry during normal operation. When commissioning furnaces, refractories need to be heated slowly to expel residual water and help lengthen the heater's life. Special steam heating and drying protocols should be used on the fuel-gas lines before lighting the pilot burners and raising the burner temperatures. General guidance for catalyst loading includes low-density and high-density catalyst systems. Tightness tests are used to confirm that process units handling hazardous materials do not leak. Vacuum systems must be leak tested, as well. The final piece of guidance focuses on Nitrogen inerting systems, describing different methods such as evacuating with steam ejectors or steaming out the system and then introducing the Nitrogen, pressurizing and depressurizing or sweeping with Nitrogen.

Simultaneous Operations (SIMOPS), if applicable, should be addressed during the steps of the commissioning and initial start-up efforts. Note that the plans for safe and efficient operations and maintenance *once the equipment and processes have been commissioned* should address the facility's ongoing Environmental, Occupational Health and Safety (EHS), and process safety management system, as well. These management systems should include policies and procedures based on the RBPS elements, such as [14]:

- Process knowledge and technologies, including new or updated equipment files, Piping and Instrumentation Diagrams (P&IDs), Process Flow Diagrams (PFDs), etc.
- Operating procedures, including normal start-ups and shut-downs;
- Safe work practices for maintenance-related and specific, non-routine activities;
- An asset integrity and reliability program, such as an Inspection, Testing, and Preventive Maintenance (ITPM) program for the new equipment (including purchasing and warehousing support personnel);
- A contractor management program to ensure that qualified contractors execute their activities safely and competently;
- Training programs to ensure performance goals are achieved (including safely operating and maintaining the equipment);
- Managing change procedures, to ensure that the hazards and risks associated with changes to raw materials, equipment, processes, and personnel are reviewed and approved before being implemented;
- Emergency management systems to ensure safe emergency response and to help reduce the consequences of loss events, including internal emergency response teams and

procedures and an Emergency Response Plan (ERP) that may
include external resources as appropriate;

- Incident investigation procedures to detect for abnormal
  situations or respond to loss events to identify and
  understand gaps and then make and implement continuous
  improvement recommendations; and
- Measurement and metrics, such as identified Key
  Performance Indicators (KPIs) for the operating processes.

Note that most of these risk management systems' content, such as
the EHS and process safety procedures, should already exist at a
brownfield site.

If the project involves a major expansion or new facility using
existing technologies, experienced operators from other locations
should be included to help support or take the lead on the
commissioning and initial start-up activities. Sometimes the local
recruited personnel may have little or no operating, maintenance, or
process safety experience (especially for greenfield projects located in
undeveloped, rural areas with little industry). Thus, it is important to
include these new employees in the commissioning and start-up
activities as much as possible. This initial start-up experience will
complement their classroom training, which focused on general
equipment-specific instructions the operating or maintenance
procedures, and field tours, since they will be able to observe and learn
from more experienced personnel during the transition time. Refer to
other resources for details on training and competence assurance
during these times of increased administrative controls [14] [21] [40]
[88] [89].

9.3.2   When managing changes to equipment or process units

Smaller projects, such as the MOC- or maintenance-related projects
described in Chapter 4, need to have the same disciplined project
management approach that the larger projects have during the

commissioning and initial start-up stages. Although there may be fewer members and groups associated with the commissioning team, robust and clear handovers should be established for safe start-ups. At this point, it is worth recognizing that no start-up-related pre-plans can anticipate all situations that actually occur during the start-up. Some general guidance for effectively managing these unexpected situations is provided in the Appendix.

9.3.3   When managing new equipment or process units

Larger capital projects will range from new, major processing equipment, such as a distillation tower, to a process unit or facility. As was noted in Chapter 5, these projects need to have the rigorous and disciplined project management approach for their commissioning and initial start-up stages. Due to the larger number of team members and groups associated with the commissioning and initial start-up efforts, robust and clear handovers should be established for safe start-ups. In such cases, a special commissioning team with a commissioning project manager should be used to help manage the risks associated with the many inspections, tests, and equipment– and process unit–related verification steps (refer to Section 9.3.1).

9.4 Incidents and lessons learned, commissioning and initial start-ups

Details of some commissioning-related incidents are included in this section. The incident summary is provided in the Appendix.

9.4.1 Incidents during commissioning and initial start-ups of changed equipment or process units.

C9.4.1-1 – Pneumatic pressure testing of pipes and vessels during initial start-up [90]

Incident Year: Three incidents, one reported in 2007

Cause of incident occurring during the initial start-up: Compressed air or nitrogen used to pressure test new equipment

Incident impact: Fatalities from the catastrophic failure of piping, flanges, and equipment that exploded during the pressure testing

Risk management system weaknesses:

LL1) Pneumatic pressure testing using high-energy compressed gases can cause catastrophic equipment failure. When feasible, hydrostatic water pressure testing is recommended: water is a non-compressible fluid, containing less energy than a compressed gas. If pneumatic pressure testing should be performed, safe work permits and a thorough safety review will help ensure that personnel are not at risk when the equipment is being tested.

Relevant RBPS Elements:
- Hazard Identification and Risk Analysis
- Safe Work Practices

C9.4.1-2 – Adding filters in new piping to collect construction-related residues [2, p. 50]

Incident Year: Before 2015

Cause of incident occurring during the initial start-up: Air introduced to piping during the installation of filters before a flammable gas compressor to prevent dirt from entering the compressor

Incident impact: Air reacted with the flammable gases in the piping system, creating decomposition products further downstream and a major fire that caused major delays to the unit start-up

Risk management system weaknesses:

LL1) Although there were many details reviewed before the start-up, creative solutions created during the time expected for resuming operations per the pre-planned start-up schedule introduced an unaddressed hazard. The changes were not adequately reviewed.

Relevant RBPS Elements:

- Hazard Identification and Risk Analysis
- Management of Change

9.4.2   Incidents during commissioning and initial start-ups of new process units or facilities

C9.4.2-1   ConAgra and Kleen Energy [91] [92] 87]

Incident Year: 2009 and 2010

Cause of the incidents occurring during the start-up: Planned commissioning activities led to a large release of flammable natural gas when workers and ignition sources were in the confined area.

Incident impact: The natural gas explosion at ConAgra caused four fatalities and injured 67 people. The natural gas explosion at Kleen Energy caused six fatalities, injured almost 50 more people, and severely damaged the facility under construction.

Risk management system weaknesses (LL — Lessons Learned for improvements):

LL1) The standard industry practice at the time was to use the natural gas in a high pressure, high volume "gas blow" through the piping to force debris out of the piping, a part of the commissioning phase of a project. The natural gas and debris are subsequently released directly to the atmosphere. Since then, this practice has been highly discouraged due to the potential for fatalities, injuries, and property damage. If a flammable gas should be used for the gas blow, then a thorough job safety review should be performed.

Relevant RBPS Elements:

- Hazard Identification and Risk Analysis
- Safe Work Practices

C9.4.2-2 – Batch processing computer programming issues during initial facility start-ups [93]

Incident Year: Before 2006

Cause of incidents occurring during the initial start-up: Computer programming and wiring issues during design and construction, respectively, of highly automated facilities

Incident impact: No serious process safety consequences, although all cost some money and lost productivity

Risk management system weaknesses:

LL1) Programming steps for batch operations were not tested adequately during pre-commissioning activities, such that specific cross-verification steps, or inherent operator "common sense," were missed during engineering design when specifying the computer coding requirements and, in a separate incident, when installing hard wiring in the field

Relevant RBPS Elements:

- Process Safety Competency
- Process Knowledge Management
- Operational Readiness

C9.4.2-3 – Small caustic leak issue upon new refinery start-up [94]

Incident Year: 2012

Cause of the incident occurring during the initial start-up: During a longer-than-expected repair time for a leak discovered during a new refinery start-up, undiluted caustic continued to be added to the crude already charged to and circulating in the partially shut-down system (a "warm circulation"). The undiluted caustic in the system upon full restart vaporized as the temperatures increased, corroded thousands of feet of stainless steel pipe, fouled almost 50 heat exchangers, and damaged instrumentation, the distillation tower, and components in the furnace.

Incident impact: Soon after running the refinery at its normal elevated temperatures and pressures, a series of quickly-extinguished fires on the new pipeline occurred, and then a heater ruptured once crude flow was resumed. Accelerated corrosion caused significant pipeline and equipment damage and a subsequent significant delay in and cost of the refinery start-up.

Risk management system weaknesses:

LL1) *Note:* No formal incident investigation report has been made publically available.

The following issues may have contributed to this incident: 1) the unanticipated delay in fixing the a leak (this was not recognized as a change in the planned start-up); and 2) the unanticipated effect of continuing to add caustic to the small amount of crude still in the unit during the warm circulation (either from a failed valve on the caustic system or by not shutting the caustic addition system off).

Relevant RBPS Elements:

- Process Knowledge Management
- Hazard Identification and Risk Analysis
- Management of Change
- Operational Readiness

9.5 End-of-life shut-down considerations

The shut-down for an end-of-life project is the transition time between the end state of a normal operations (the equipment or process unit has been shut-down using normal shut-down procedures), and meeting the equipment's temporary end state (i.e., being mothballed) or its final end state (i.e., being decommissioned). These choices for the equipment end state were illustrated in Figure 9.2. The next three sections briefly highlight some specific end-of-life decommissioning planning considerations.

9.5.1   Structural engineering surveys

An engineering survey, conducted by a competent, contracted surveyor, is used to identify potential hazards and to evaluate the condition of structures and buildings thoroughly before the decommissioning activities are designed and executed. These structural surveys will identify structural-related hazards that could pose additional hazards during the decommissioning activities. Thus, this survey will help evaluate the potential risks of an unplanned structural collapse and the impact of the decommissioning activities, such as deconstruction and demolition, on surrounding equipment or structures. This survey should include documentation of damage to the structures (which may affect structural integrity), methods and safeguards to prevent harm to the personnel executing the tasks, and risks to other assets. Challenges for the surveyor may include locating the original construction and structural drawings, typically due to inadequate or non-existent document control and management of change systems. Additional details, including the typical content for an engineering survey report are provided elsewhere [31, pp. 252, Table 11.1].

9.5.2   Planning for decommissioning projects

Primary concerns that should be considered when planning for decommissioning include de-inventorying the plant of hazardous materials and managing the structural integrity during dismantling and removal. Since incidents have occurred during the decommissioning times associated with equipment– and process unit–related shut-down activities, the following issues should be considered for potentially unknown factors that may affect process safety [31, p. 250]:

- Changes from the facility's design introduced during construction that may or may not have been approved (i.e., weaknesses in the project change management system);
- Subsequent modifications that altered the original design that may or may not have been approved (i.e., weaknesses in the facility management of change system);
- De-inventorying and disposing of all process fluids, catalysts, and other materials;
- Residual hazardous materials within process vessels, piping, insulation, and structural members, such as process chemicals, asbestos, lead, heavy metals, etc., requiring special handling;
- Unknown strengths or weaknesses of construction materials due to aging; and
- Hazards created by the tasks necessary for the decommissioning methods used (e.g. hazards to adjacent process units, third party installations, and surrounding facilities or communities).

In addition, these decommissioning tasks may include, but are not, limited to [31, p. 250]:

- Handling, storage, and disposal of hazardous materials;
- Handling and storage of explosives;
- Removal of heavy equipment and structures;
- Integrity of partial and/or damaged structures;

- Working at heights;
- Working near or over water;
- Presence and/or removal of overhead/underground/subsea pipelines and utilities; and
- Location of temporary facilities, (e.g. trailers) associated with deconstruction activities.

If the end-of-life project is for demolition or deconstruction, potential issues may involve, but are not limited to [31, p. 24]:

- Proximity of neighboring facilities and buildings may require dismantling and prohibit toppling/explosives,
- Deconstruction of some equipment for future re-use,
- Partial decommissioning of operating facility,
- Presence of asbestos and PCBs in older facilities,
- Simultaneous operations with adjacent facilities,
- Vibration that might impact adjacent operations,
- Underground cables and piping, and sewers, including unknown
  - Locations, and
  - Connections to other adjacent facilities,
- Environmental remediation.

These issues require careful planning to perform decommissioning safely and efficiently, and personnel should be fully knowledgeable of the hazards and the appropriate safety measures selected to mitigate the hazards and reduce the activity's risks. Some companies may perform a formal "Hazards of Decommissioning (HAZDEM)" review before decommissioning, as well, to help ensure that the decommissioning activities are incident-free. When compared to deconstruction and dismantling activities, it is important to recognize that there may be greater hazards to personnel or other equipment during demolition activities. Proper planning and strong oversight and guidance with a decommissioning team project manager is essential to avoid injuries and incidents.

Note that these risk management and the other process safety considerations, if considered at the concept and engineering design stages of a project, would significantly help reduce the risks during decommissioning. Thus, when planning and developing a project, especially in projects making major equipment or process unit changes (including greenfield projects), it is important that the equipment, process unit, and facility designs: include provisions for their eventual decommissioning; and address the initial siting and layout of the equipment, process units, and facilities for their eventual decommissioning [67].

9.6 Mothballing considerations

When equipment or process units are taken out of service with the possibility of future use, special decommissioning procedures may include depressurization, de-inventorying and cleaning, and additional equipment-specific preservation measures. Mothballing projects may include abandoned-in-place equipment when it is uncertain how long the equipment will be inactive, not used, and idle. Ongoing testing or preventive inspection tasks may need to be performed to maintain the equipment integrity so it can be operated safely once it is returned to active service.

Depending on the types of hazards associated with the process materials or the equipment's construction materials, these measures may include pickling or a dry nitrogen purge to maintain a proper internal atmosphere to help prevent corrosion. For example, iron sulfide, a pyrophoric compound that ignites when exposed to oxygen, is formed when unremoved hydrogen sulfide reacts with rust, which could have formed inside the mothballed equipment. Moreover, note that if inert atmospheres are used, the mothballed equipment should be labelled to warn the personnel of unsafe atmospheres inside the equipment.

If there is an area where mothballed equipment items will be stored for possible use later, there will be similar asset integrity management challenges, including retaining the specific equipment design and inspection documentation for the equipment located in the equipment "boneyard." If the mothballed equipment or process unit does start-up later, special recommissioning procedures should be prepared to reverse any preventive procedures that were used when preparing the mothballed equipment.

It is important that all ITPM tasks are completed and verified before restart. In addition, these Inspection, testing, and preventive maintenance tasks should be re-entered in the facility's ITPM programming schedule to ensure ongoing asset integrity. The facility or equipment should also be subject to an operational readiness review prior to restart [14]. At some point, the mothballed and boneyard equipment will eventually progress to the permanent decommissioning stage (see Section 9.8).

9.7 Incidents and lessons learned, mothballing

Details of some mothballing-related incidents are included in this section. The incident summary is provided in the Appendix.

9.7.1   Incidents when recommissioning mothballed equipment

C9.7.1-1 – Starting up a Mothballed Vessel [20, p. 455]

Cause of incident occurring during the start-up: Unremoved residue reacted with the fresh ingredients.

Incident impact: The reaction caused an increase in temperature and a gas release into the working area.

Risk management system weaknesses:

LL1) Recommendations from the incident investigation included completely emptying equipment if being "mothballed" (out of service longer than usual), testing any residual material in the equipment before reusing it, and prevent deterioration of any residual residues with a protective barrier (in this case water or other solvent).

Relevant RBPS Elements

- Process Knowledge Management
- Hazard Identification and Risk Analysis
- Operating Procedures
- Operational Readiness

9.7.2   Incidents when executing decommissioning efforts on mothballed equipment

At the time of this guideline's publication, no incidents had been identified which occurred during this transient operating mode.

9.8 Decommissioning considerations

Decommission consideration will depend on the scope of the project and facility-specific factors. As mentioned in Section 9.2, robust handovers, especially those involving contractors, are essential to prevent incidents (also refer to Table 9.2). This includes the demolition activities, which are typically carried out by an outside demolition contractor. These specialized contractors should be qualified to operate the required demolition equipment, and, most importantly, they should have training on the hazards of the chemicals associated with the decommissioned equipment and the knowledge and ability to handle them (especially if the equipment was not adequately cleaned).

In addition, these contractors should understand that all unanticipated situations should be communicated to facility personnel, and any changes in the accepted and reviewed demolition's scope should be reviewed and approved again by the company before any additional work is performed. Decommissioning incidents have occurred when the scope changed and work was performed without an adequate review. A detailed example of a decommissioning checklist is provided elsewhere [31, pp. 515, Appendix E].

9.9 Incidents and lessons learned, decommissioning

Details of some decommissioning-related incidents are included in this section. The incident summary is provided in the Appendix.

9.9.1   Incidents during equipment decommissioning

C9.9.2-1 Redundant Piping Removal [20, pp. 367-368]

Incident Year: Not known

Causes of the incident occurring during the decommissioning: Wrong joint broken during activity to remove redundant, unused pipes.

Incident impact: Release of 7 bar (100 psig) compressed air; mechanic escaped injury after breaking the flange.

Risk management system weaknesses:

LL1) The incident description noted the following: An ineffective permit-to-work system that did not address Safe Work Practices, such as labeling of the piping that was to be removed and physical locks to prevent the wrong flange from being opened.

Relevant RBPS Elements:

- All twenty elements in the CCPS RBPS model.

9.9.2   Incidents during process unit or facility decommissioning

C9.9.2-1   Bhopal [21] [95] [96] [97]

Incident Year 1984 (*Note:* CCPS was created in 1985 due to several international companies response to Bhopal; CCPS RBPS was issued in 2007)

Causes of the incident occurring during the decommissioning: Lack of understanding of "process safety" as is known today, including safety leadership and culture, operational discipline, and process safety systems.

Incident impact: A release of 40 metric tons of a toxic chemical that caused over 3,600 fatalities documented (reports differed on actual numbers, as over 16,000 fatalities were claimed), more than 100,000 injuries, significant harm to livestock and crops surrounding the facility, and the ultimate break-up and elimination of the parent company.

Risk management system weaknesses:

LL1) Lack of understanding the three foundations in process safety: process safety leadership and culture, operational discipline, and process safety systems.

Relevant RBPS Elements:

•   All twenty elements in the CCPS RBPS model.

9.10   How the RBPS elements apply

All Risk Based Process Safety elements (RBPS) apply to initial start-up and decommissioning efforts. The initial and final life cycle steps for these efforts occur once in the lifetime of the equipment or process. This is a major difference between these life cycle steps projects and those from the other start-ups or shut-downs described in earlier chapters, which occur many times during the use life cycle.

10 Risk Based Process Safety (RBPS) Considerations

10.1 Introduction

This chapter provides a brief overview of the four pillars and their corresponding elements in the CCPS Risk Based Process Safety (RBPS) model (Section 10.2), as is illustrated with the RBPS model in Figure 10.1 (Adapted from [98]), and provides a summary of how the RBPS elements apply to each of the transient operating modes (Section 10.3).

The RBPS Model presents an overall process safety and risk management program with several tightly-coupled elements. For example, a company cannot develop an effective asset integrity program if it does not have a thorough technical understanding of design specifications. For example, the construction materials used to fabricate the equipment will affect how the equipment degrades when it is in service. The next few sections discuss how weaknesses in one or more of these elements contributed to a less effective process safety management program and adversely impacted the organization and its ability to manage its transient operating modes safely.

There was a pattern for some pillars and elements during the transient operating modes to be more prevalent than others as factors contributing to an incident. These weaker elements did not exist or were not implemented effectively at the time of the incident. This chapter provides some insights on the adverse process safety performance effects from the weaker elements in these pillars as follows:

- Effects from a weak commit to process safety pillar (Section 10.3.1)
- Effects from a weak understanding of the hazards and risks pillar (Section 10.3.2)

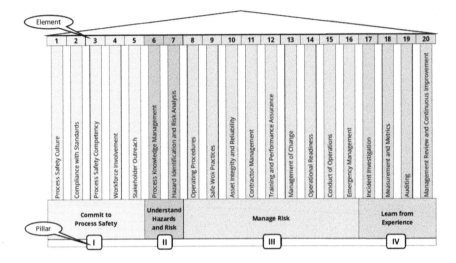

Figure 10.1 CCPS Risk Based Process Safety (RBPS) Model.

- Effects from a weak application of the risk management pillar (Section 10.3.3)
- Additional reflections on the asset integrity and reliability element (Section 10.3.4)
- Effects from a weak learning from experiences pillar (Section 10.3.5)

Since a strong safety culture and steadfast operational discipline are essential for effectively managing a process safety program, their effect on the overall business risk, especially during the transient operating mode, is covered separately in Section 10.4. This chapter ends with some insights on how process safety performance can be improved (Section 10.5). Learning and sharing with others is the best way to improve process safety performance.

10.2   An RBPS Overview

10.2.1 Pillar I: Commit to Process Safety

The "Commit to Process Safety" pillar is the cornerstone of process safety excellence. A workforce that is actively involved and an organization that fully supports process safety as a core value will tend to do the right things in the right ways at the right times—even when no one else is looking. The five elements within with this pillar are:

Element 1—Process Safety Culture: A committed, positive process safety culture provides a nurturing environment where employees at all levels can effectively manage process safety. This culture begins at the highest levels of the company, influences decision-making across the organization, and is shared by everyone in the organization.

Element 2—Compliance with Standards: A committed company complies with expected regulations, directives, standards, and codes issued by governmental jurisdictions, consensus standards groups, and the company. Progressive companies use the latest technologies and methods that often go beyond the minimum compliance requirements.

Element 3—Process Safety Competency: A company's competency contains three aspects of process safety knowledge and information: 1) continuing to learn and improve; 2) transferring information effectively within and in between groups; and 3) applying the information. Everyone effectively manages their process hazards and risks due to their specific skills or access to resources with the skills.

Element 4—Workforce Involvement: A company's frontline workforce—including supervisors, operators, mechanics, electricians, and contractors who work day-to-day with the equipment and the hazardous materials or energies—can and will contribute to the success of safe operations when they are involved in developing and sustaining parts of the RBPS elements.

Element 5—Stakeholder Outreach: The company benefits from dialogues with the public, emergency responders, professional groups, and regulatory agencies by sharing information on the plant's process hazards, the methods used to safely manage risks, and how to respond safely if a loss event occurs.

10.2.2 Pillar II: Understand Hazards and Risk"

The "Understand Hazards and Risk pillar is the technical foundation of a risk-based approach, establishing the technology framework that is essential when developing how the risks are managed (Pillar III) and monitored (Pillar IV). Everyone should understand what the process hazards are and how the risks have been assessed. The two elements within this pillar are:

Element 6—Process Knowledge Management: This element focuses on the recorded information, that includes verified, accurate, up-to-date, technical documents, engineering drawings and calculations, and specifications used to fabricate and install the process equipment. This information is accessible to everyone working with the hazardous materials and energies.

Element 7—Hazard Identification and Risk Analysis (HIRA): The technical information is used in this element to help identify the process hazards and their potential consequences. Each company has a risk tolerance level, helping each plant assess their risks using qualitative or quantitative approaches. The risk analyses are used to establish the engineering and administrative controls needed to help safely manage the risks.

10.2.3 Pillar III: Manage Risk

The "Manage Risk" pillar contains the elements needed for safe operations, using and *applying* the technical knowledge and risk analyses from Pillar II. Once the process hazards and risks are understood, their engineering and administrative controls should be implemented and sustained when the process is operating. The nine elements associated with this pillar are:

Element 8—Operating Procedures: These procedures are documented (written) instructions for the manufacturing operations that describe how the operation is to be carried out safely, explaining the consequences of deviation, describing key safeguards, and addressing special situations, such as transient operating modes and emergencies.

Element 9—Safe Work Practices (SWP): These are work practices that apply to the entire facility, whether the activity is being performed on a hazardous process or not. They bridge the gap between operating procedures (Element 8) and maintenance procedures (Element 10), including permits-to-work, line breaks, confined-space entry, hot-work permits, electrical isolation, fall protection, and construction-related activities, such as excavations or lifts.

Element 10—Asset Integrity and Reliability: These activities focus on maintenance activities, such as an Inspection, Testing, and Preventive Maintenance (ITPM) program, ensuring that the equipment remains suitable for its intended purpose. This element includes documented maintenance-specific procedures to ensure that equipment handovers between the operations and the maintenance groups clearly identify the state and potential hazards associated with the equipment.

Element 11—Contractor Management: The contractor management system helps ensure that contracted services support the operations and personnel safety performance goals. This system also addresses the selection, acquisition, use, and monitoring of the contracted services.

Element 12—Training and Performance Assurance: Training includes classroom and field instruction in job and task requirements for operating and maintaining the equipment, including process safety-related tasks and responsibilities of supervisors, engineers, leaders, and process safety professionals. The performance assurance step verifies that the knowledge-based training can be applied with competence.

Element 13—Management of Change (MOC): An effective MOC system provides a structured reviewing and authorizing protocol for proposed changes process or equipment designs, operational or maintenance procedures, and personnel. The MOC system ensures effective communications between groups during handovers, focusing on the changes to the process hazards and associated risks, and is often documented through a Pre-Startup Safety Review (PSSR).

Element 14—Operational Readiness: An Operational Readiness Review (ORR) extends beyond a PSSR performed through an MOC, ensuring beforehand that the process can be safely started-up or restarted. Thus, this element applies to the start-up of equipment or facilities and to the initial start-ups of new facilities. In addition, this element applies to start-ups after process shutdowns (e.g., small projects, maintenance), after facility shutdowns (e.g., turnarounds), and equipment idled for months, mothballed for years, or repurposed for a different process.

Element 15—Conduct of Operations: This element is driven by an operational discipline for everyone in the organization to carry out deliberately and faithfully their tasks in a safe and structured approach. It is closely tied to the organization's culture (Element 1).

Element 16—Emergency Management: The emergency management system includes planning for possible emergencies, providing resources needed to safely execute the plan, informing everyone at the plant how to report an emergency, and practicing and debriefing the plan with emergency responders (e.g., tabletop and field exercises). The plan also includes how the employees, contractors, and neighbors will be notified (e.g., alarms) and what they should do (such as when relocate to a safe shouldering location, when to evacuate, or when to shelter-in-place). Depending on the impact of the emergency, the plan also includes how the facility will communicate with local authorities and external stakeholders.

10.2.4 Pillar IV: Learn from Experience"

The fourth pillar, "Learn from Experience," focuses on the Plan-Do-Check-Act monitoring phase. Metrics provide direct feedback on the workings of RBPS systems, and leading indicators provide early warning signals on any process safety performance-related results. When companies share lessons learned from their experiences, others can learn from them, apply the learnings within their own group, and help prevent the incident from occurring again. The four elements associated with this pillar are:

Element 17—Incident Investigation: This element provides a process for identifying, reporting, tracking, and investigating incidents and near-misses to help identify root causes. The investigation system generates recommendations, ensures that corrective actions are implemented, and shares the lessons learned with others. Incident trends are evaluated in conjunction with other elements in this pillar.

Element 18—Measurement and Metrics: This element helps an organization identify leading and lagging indicators of process safety performance, including incident and near-miss rates. The metrics show how well the RBPS elements are performing, helping an organization identify gaps, implement recommendations, and improve their overall process safety program performance.

Element 19—Auditing: Process safety-related audits are structured and periodic reviews of a process safety management system's performance, and are based on standards, guidelines, and regulatory requirements. Auditors are independent, help identify gaps in performance, and issue a formal report with recommendations to address the gaps. The auditing system is then used to document and track the recommendations until closure.

Element 20—Management Review and Continuous Improvement: The management review complements the auditing element by having managers, at all levels in the organization, evaluate and review the performance of their process safety expectations and goals with their staff. These reviews are more frequent that a full process safety management system audit (Element 19) to help managers understand and address potential process safety-related issues proactively.

The next few sections discuss how weaknesses in one or more of these elements contributed to a less effective process safety management program and adversely impacted the organization and its ability to manage its transient operating modes safely. When process safety-related risks are not identified and assessed by those making decisions, no matter what level in the organization they are, process safety incidents may occur.

10.3   Applying RBPS to each transient operating mode

Although weaknesses in each of the CCPS RBPS elements contributed in many of the transient operating mode incidents, a few of the elements tended to occur more often than other elements. These RBPS elements directly or indirectly contributed to the incident due to them either not being effectively implemented, managed, or sustained at the time the incident.

The RBPS element weaknesses identified during the transient operating mode from this incident review are compiled in the Appendix and are summarized in Figure 10.2. Every incident occurred when the safeguards did not work per design or as planned.

As shown in Figure 10.2, weaknesses in eight of the twenty RBPS elements contributed to over seventy percent of the incidents during the transient operating modes. Listed by pillar, the eight contributing elements included:

- Pillar I (Pillar with 11% of the time as an identified contributor); Process Safety Competency; Process Safety Culture;

- Pillar II (24%): Hazard Identification and Risk Analysis, Process Knowledge Management;
- Pillar III (37%): Asset Integrity and Reliability, Operating Procedures, Management of Change, and Emergency Management.

This overall trend is observed: When Pillar I's elements were weak, Pillar II's elements were compromised, and when Pillar II's elements were weak, the effectiveness of the elements in Pillar III were compromised. Each of these three pillars and their eight corresponding elements, those contributing most often to incidents during transient operating modes, are discussed in more detail in the next few sections.

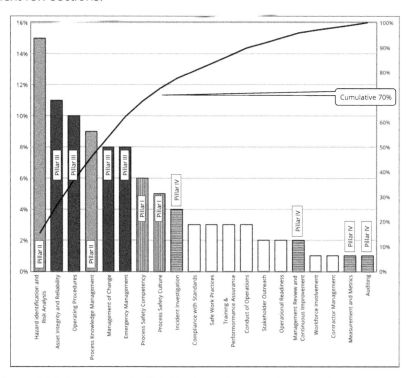

Figure 10.2 Summary of CCPS RBPS Weaknesses from Incidents Occurring During the Transient Operating Modes

Note: The incidents used to generate Figure 10.2 are provided in the Appendix.

10.3.1 Effects of a weak commit to process safety pillar

"It is simply not possible to talk about process safety without considering the impact of the safety culture and the leadership at a company..."

J. A. Klein [21, p. 31]

The transient operating mode incidents occurred, in part, to a weak leadership-driven process safety culture, the first RBPS element in the first RBPS pillar (Figure 10.1). The process safety competency element, also in pillar I, was within the top eight elements identified, as well. Strong leadership supports development of the different process safety competencies across their organization. These competencies are then applied with steadfast operational discipline.

Some additional reflections on process safety culture and leadership, Element 1: A safety culture is the "normal way things are done...reflecting expected organizational values, beliefs, and behaviors..." [21, p. xxi]. This culture is committed to process safety, sets the priorities for safe process safety performance, and provides sufficient resources in personnel, technologies, and equipment to maintain safe process safety performance. It is the collective need to prevent major incidents and to *do the right thing* [99]. Leaders at all levels in an organization that has a committed, positive process safety culture will nurture everyone, identify the hazards, evaluate the risks, manage the processes, and learn from both good and bad experiences.

A strong process safety culture and leadership instills safe operations during *all* of the transient operating modes, as well during normal, abnormal, and emergency operations. Process safety culture was noted in more transient operating mode incident reports after

2005, when it was raised as a significant factor in a refinery explosion that same year. Every incident occurred when the safeguards did not work per design or as planned, often occurring when the company's leadership did not support implementing or sustaining the critical process safety equipment or did not adequately train people supporting one or more stages of the equipment's life cycle. For this reason, additional guidance on improving a company's process safety has been published [100].

10.3.2 Effects of a weak understanding of the hazards and risks pillar

Pillar II, containing the process knowledge, hazards identification, and risk analyses elements, contributed to twenty-five percent of the transient operating mode incidents shown in Figure 10.2. When the normal start-up and shut-down procedures are developed and implemented correctly, the process safety risks during these transient operating modes are minimized. However, since every abnormal or emergency shut-down and start-up cannot be predicted, the technological process knowledge must be understood by those responding to the unpredicted event before making decisions and action. Often the lack of awareness of the potential hazards and their associated risks only became apparent after the abnormal or emergency transient operating mode incident occurred. Lack of awareness of the process hazards and risks inevitably led to the incident.

10.3.3 Effects of a weak application of the risk management pillar

These four Pillar II elements, in this order, contributed to 37% of the transient operating mode incidents: Operating Procedures, Asset Integrity and Reliability, Management of Change, and Emergency Management (Figure 10.2). Without the understanding of the process safety risks evaluated and minimized through the risk reduction efforts in Pillar II, the elements in Pillar III could not be designed, implemented, or sustained effectively. Thus, these risk management programs and systems were inadequate when responding to the transient operating mode shut-downs caused abnormal or emergency operations.

Since administrative controls are less effective than engineering controls, it should come as no surprise that inadequate Operating Procedures during the transient operating mode rank as the weakest of all the elements (refer to the hierarchy of controls [21, p. Figure 3.6]; and Figure 10.2). Often, this was due, in part, to unwritten, quick, and inadequately reviewed decisions and actions taking during start-up upsets. The start-up procedures simply cannot address all unknowns, with the responses (or lack thereof) sometimes proving fatal. The established normal start-up and shut-down procedures should guide decisions for safely addressing unexpected issues as they arise (see the Appendix for more unexpected shut-down-related guidance).

Unexpected start-up issues occurring after a project, maintenance, or facility shutdown were often caused by an unexpected start-up condition of the equipment. At other times, abnormal operation upsets may have caused an unusual equipment idling condition or an emergency shut-down Again, incidents during start-ups occurred because the equipment was not in its normal start-up condition. The lack of understanding the state of the equipment could have been due, in part, to inadequate handovers. Refer to Chapter 3 for a more detailed discussion on preventing inadequate procedural handovers.

The second most noted Pillar III element contributing to transient operating mode incidents was the Asset Integrity and Reliability element. If the equipment failed catastrophically during normal operations, operations had to implement an immediate emergency shut-down. Also, when the safeguarding equipment failed to operate as expected during the abnormal or emergency operation, the consequences of the incident were worse. The process and safeguarding equipment failures could have been prevented with a robust Inspection, Testing, and Preventive Maintenance (ITPM) program.

The Management of Change element was the next weakest Pillar III element that contributed to the transient operating mode incidents (Figure 10.2). Formally implemented and sustained management of change programs have helped improve the handovers between groups executing process or facility shutdowns. However, companies have struggled with effectively managing changes that occur when things do not go as planned. Such unexpected changes occurred during the commissioning and initial start-up stages. Thus, the pre-start-up safety or the operational readiness reviews are designed and performed to adequately ensure smooth handovers. However, an unexpected equipment condition or response led to troubleshooting. The results of the troubleshooting efforts includes temporary shutdowns or placing the equipment in a "hold" position for a much-longer-than-expected time.

Emergency management issues, the sixth weakest element contributing to incidents during the transient operating mode arose during emergency shut-downs. The shut-down incidents occurred when either the administrative and engineering controls were inadequate. Personnel could not safely respond to the emergency or the safeguarding equipment did not exist or function as planned.

Some of the transient operating mode incident reports noted that personnel were unable to find the guidance they needed to safely respond to or even manage the loss event. Sometimes people did not know how to use the procedures, including how to respond safely, lead the response, and account for site personnel. Sometimes the communications within the company hierarchy, to external emergency responders, and to the affected communities surrounding the facility did not exist or were not understood and implemented. An effective emergency management program includes regular emergency response plan drills and debriefing sessions. Additional emergency management guidance is provided in other publications [14, p. Chapter 18] [21, p. Chapter 11].

10.3.4 Additional reflections on the asset integrity and reliability element

As noted in Section 10.3.3, some incidents occurred during the transient operating mode were due, in part, to weaknesses in the asset integrity and reliability program. Many of these incidents occurred due to compromised equipment that did not perform as expected—the equipment was not fit for use. No matter when the equipment failure occurred, the failure of equipment that was properly designed, fabricated, installed, and operated is due to it not being *maintained* through a robust ITPM program on both the preventive and mitigative engineering controls. (More discussion on the overall equipment life cycle was provided in Chapter 9.) Strong leadership support of resources to manage ITPMs is crucial, as inadequate maintenance can be attributed to and have led to some of the worst industrial incidents, with some noted specifically during start-ups due, in part, to unreasonable maintenance schedules [101, pp. 124-125].

Unfortunately, some incident reports noted that the Process Hazard Analysis (PHA) had simply "failed to identify" the causes of the incident, and thus, the "unexpected" or unanticipated failure occurred and there were insufficient safeguards in place to protect personnel, the environment, or property from harm. Since several incident reports noted the HIRA as a weak element without considering what a PHA Team should assume during its review, this element appeared to be a much larger contributor than it probably is. This viewpoint tended to skew the results more to the ineffective HIRA RBPS element as shown in Figure 10.2. No PHA team can anticipate *every combination* of potential failures that could foreseeably occur in a complex process. Thus, the prudent PHA approach used in Element 7 *assumes* that all the other elements in Pillar III have been adequately designed, implemented, and *are being sustained*. This includes assuming that a robust ITPM program exists and is effective. Thus, "unexpected" equipment failures should not occur. A more detailed discussion of the types of assumptions made by a PHA HAZOP Team are provided in the Appendix.

As was discussed in Chapter 4, the maintenance strategy for process shutdowns includes coordinating the schedules between the operations and maintenance groups, as the equipment should be available when its maintenance is performed. Inspections and tests which do not require time for a process unit or utilities shutdown time are easier to schedule and perform; those that do require a shutdown time should be coordinated with the production schedulers. Emergency repairs on equipment that failed unexpectedly delay the expected production schedule, adding undue stress to personnel across the organization. Again, a robust ITPM program should be in place to help prevent unexpected failures, especially for critical equipment used in the engineering controls required for safe operations.

One approach to ensure a robust ITPM program poses two questions that help identify critical equipment: How are the essential controls identified and how are the controls maintained to prevent unexpected failure? These questions apply to all operating modes: normal, abnormal, emergency, and transient. The answers to these questions follow.

1) How are the essential controls identified? The critical engineering and administrative controls can be identified through a Process Hazard Analysis (PHA), often by selecting high-risk scenarios from a Hazards and Operability Study (HAZOP) and performing a Layer of Protection

Analysis (LOPA) on them (RBPS Element 7, [102]). The HAZOP/LOPA approach, described briefly in Chapter 8, can be used to identify Independent Protection Layers (IPLs). The IPLs identified as engineering controls are the preventive and mitigative equipment essential for operating the process safely.

2) How are the controls maintained to prevent unexpected failure? After the PHA team's effort is complete, the maintenance group can use the PHA's IPL equipment list to identify critical equipment that should be on the ITPM program. The maintenance group understands the equipment's failure mechanisms in the field and implements monitoring approaches that help anticipate equipment degradation or pending failure. A brief discussion on how to monitor equipment integrity was provided in Chapter 6. The abnormal situation management approach for equipment condition includes both engineering controls (i.e., predictive monitoring and response) and administrative controls (i.e., alarm management and response). Additional resources specific to asset integrity efforts are provided elsewhere [23] [60].

10.3.5 Effects of a weak learning from experience pillar

"Lessons are learned from such [process safety] events. But not everyone learns. Many of those who do learn forget. Or worse, they decide that the lessons don't apply to them."

P. R. Robinson [101, p. 123]

Although the elements in the learning from experience pillar accounted for only 8 percent of the incidents shown in Figure 10.2, a robust process safety program has this pillar in place for its continuous improvement efforts.

The learning from experience pillar contains these four elements: Incident Investigation, Management and Metrics, Auditing, and Management Review and Continuous Improvement. Incident investigations are reactive—the harm has been done. On the other hand, reviews and audits are proactive—identifying issues that can be addressed before harm has occurred. Although the CCPS guidance for identifying metrics using leading and lagging indicators does not specifically address indicators for the transient operating mode [103], many of the indicators used for abnormal operations would apply to transient operating modes. Potential indicators based on discussions throughout this guideline could focus on these items during the start-ups or shut-downs:

- Basic Process Control Systems (BPCS): number of times that the expected transition conditions were not met during the transient operation;
- Instrumentation and Alarms: number of times expected values during the transition time were exceeded during the transient operation;
- Start-up and Shut-down procedures: number of procedures reviewed or updated on time per the established schedule; number of times procedure had to be changed during the transition time;
- Emergency Shut-down procedures: number of procedures reviewed or updated on time per the established schedule; number of times procedure actually used during review period (typically indicates issues from other RBPS elements, such as a inadequate ITPM due to catastrophic equipment failures).

If selected indicators and metrics do not meet the expectations established for the safe start-ups or shut-downs, then proactive measures can be identified and implemented before issues arise during the transition.

This section discussed the weaker pillars and elements, which are influenced by the process safety culture and how much leadership supports and resources the process safety program across the organization. A strong process safety culture permeates into good operational discipline (discussed next), better design and implementation of effective process safety systems, and improved process safety performance.

10.4   Effects of weak operational discipline

One of the foundations of an effective process safety program, closely linked to its commitment to process safety pillar and to sustaining its management systems in Pillars II, III, and IV, is the organization's Operational Discipline (OD). A weak Conduct of Operations element (Pillar III, Element 15) is reflected by weaknesses in the company's OD. Although OD is difficult to measure, it is useful to think of its impact on risk by expressing its qualitative effect using the following simplified risk equation:

Details on this equation are provided in other publications [21, p. 85] [49]. As an example of how OD qualitatively affects the risk, the OD term is in the denominator:

$$\text{Risk} = \frac{\text{Frequency x Consequence}}{\text{f (Operational Discipline)}} = \frac{F \times C}{OD} \qquad \text{Equation 10.1}$$

The values of the denominator of Equation 10.1 indicate the following:

- 1/1 (or "1") representing 100% OD, where everyone does everything right every time, or
- 1/2 (or "0.5") representing 50% OD, where everything is done correctly—or incorrectly—*only half of the time*.

Thus, the *actual* risk at 50% OD is twice the *expected* risk with 100% OD. As OD performance increases, the closer it approaches 100% compliance and effective conduct of operations for everyone in the organization. When everyone adheres to their procedures, standards, and guidelines, especially when designing, operating, maintaining, and changing the equipment, the process safety risk is minimized.

Some examples of good OD—and how they impact the transient operating mode—for process design, hazards identification, and risk evaluations are shown in Table 10.1; for operations and maintenance in Table 10.2; for changes, emergency response, and incident investigations in Table 10.3; and for monitoring effectiveness of the process safety systems in Table 10.4.

Although OD cannot adequately represent the complexities inherent when managing process safety risks, weak OD can contribute to a company's weak process safety performance. In addition, if leadership does not have the operational discipline to implement improvements, the risk to the company will increase as the designed and implemented systems are not sustained and degrade over time. Since it is beyond the scope of this guideline to cover operational discipline in detail, the reader is encouraged to review other publications [21, p. Chapter 4] [104] [49].

10.5   Approach for improving process safety performance

Process Safety is a global, multi-industry, multi-cultural journey [105] [106]. Today's global approach has better tools for collaboration, and continues to improve every year with new information shared industry-wide in process safety-related conferences and publications. Establishing the CCPS RBPS elements in 2007 provided a foundation from which the current systemic approaches are being addressed. These elements capture many of the essential parts of an effective process safety program; how they impact each other—how the established systems are used to manage them—is a field ripe with growth and understanding, especially during transition times.

Professional organizations are collaborating regionally, as well as globally, to continue improving process safety performance, including the CCPS's "Vision 20/20" [107].

It is important to re-emphasize that people should learn from incidents via thorough, well-communicated reports of the incident lessons and directions for future process safety performance.

Our journey continues!

Table 10.1 Examples of good operational discipline for process design, hazards identification, and risk evaluations.
(Adapted from [49, p. Table 1]*)*

Process Safety System	RBPS Element	Examples of Good Operational Discipline During Transient Operations
Design Safe Processes	Process Knowledge Management	Understanding potential hazards; effectively documenting process safety knowledge and technical information used to set transition-related parameter guidelines
Identify and Assess Process Hazards	Hazard Identification and Risk Analysis	Identifying any additional hazardous conditions which may occur (e.g., temperature, pressure, or processing extremes)
Evaluate and Manage Process Risks		Defining "tolerable" risk parameters which may differ from those during normal operations; identifying additional administrative or engineering controls which may need to be implemented during the transition times

Table 10.2 Examples of good operational discipline for operations and maintenance.
(Adapted from [49, p. Table 1])

Process Safety System	RBPS Element	Examples of Good Operational Discipline During Transient Operations
Operate Safe Processes	Operating Procedures	Resourcing the operating facility adequately during the transition time; writing safe start-up, shut-down, and emergency shut-down procedures, including defining safe operating limits during the transition times; following these safe operating procedures (e.g., do not exceed safe operating limits); maintaining procedures; ensuring shift-to-shift consistency; and reviewing and validating critical procedures
	Safe Work Practices	Following safe work procedures (e.g., permit to work; job safety analyses; hot work, electrical isolation, etc.).
	Training and Performance Assurance	Defining position qualifications for the start-up and shut-down times; testing for theoretical understanding during the transition times; qualifying through skills demonstration during the transition time; and scheduling essential refresher training
	Contractor Management	Using qualified contractors (e.g., to install equipment or to inspect and refurbish/maintain the safeguards)
Maintain Process Integrity and Reliability	Asset Integrity and Reliability	Ensuring an effective Inspection, Testing and Preventive Maintenance (ITPM) program for all safeguards; resourcing maintenance teams adequately in preparation for planned or extended shutdowns; adhering to the scheduling planned maintenance on critical safeguards; using qualified personnel to inspect, test, and refurbish safeguards, as needed before start-ups or commissioning

Table 10.3 Examples of good operational discipline for changes, emergency response, and incident investigations. (Adapted from [49, p. Table 1])

Process Safety System	RBPS Element	Examples of Good Operational Discipline During Transient Operations
Change Processes Safely	Management of Change	Identifying and responding quickly to proposed changes during the transition times; evaluating the risks associated with the change; properly approving and authorizing changes; effectively communicating changes; and effectively documenting changes
	Operational Readiness	Following and completing safe work procedures before resuming operations (e.g., permit to work; job safety analyses; hot work, electrical isolation, etc.); performing operational readiness / pre-start-up safety reviews; and effectively managing and documenting handovers
Manage Incident Response and Investigation	Emergency Management	Creating contingencies to the emergency response plan if events occur during transition times; resourcing response teams adequately if special conditions need to be addressed during the transition time
	Incident Investigation	Identifying and evaluating incidents, including near misses, that occur during the transition times; resourcing investigation teams adequately; identifying causal factors (e.g., incipient and latent; systemic); identifying systemic or cultural issues; proactively evaluating unexpected events; sharing and implementing learnings from incidents at other locations (within and external to company)

Table 10.4 Examples of good operational discipline for monitoring effectiveness.

(Adapted from [49, p. Table 1])

Process Safety System	RBPS Element	Examples of Good Operational Discipline During Transient Operations
Monitor Process Safety Program Effectiveness	Measurement and Metrics	Identifying, tending, and tracking leading and lagging indicators for the transition times; addressing and sharing the findings (within and external to company, as appropriate)
	Auditing	Scheduling audits that include the transient operating mode procedures; documenting, addressing, and sharing findings (within and external to company, as appropriate)
	Management Review and Continuous Improvement	Scheduling area management reviews during the transition times; documenting, addressing, and sharing findings (within and external to company, as appropriate)

Appendix

Transient operating modes: incident review and guidance

A.1 Introduction

This appendix provides more details for the incidents used in this book and provides some guidance on anticipating deviations from normal operations when unexpected issues arose during start-ups and shut-downs. The two sections in this appendix are as follows:

- Review of incidents during transient operating modes (Section A.2)
- Managing the unexpected during transient operating modes (Section A.3)
- Anticipating and addressing unplanned issues with
  - Loss of utilities (Section A.3.1)
  - Engineering controls (Section A.3.2)
  - Administrative controls (Section A.3.3)

A.2 Review of incidents during transient operating modes

The incidents occurring during the start-up and shut-down times are noted in Figure A.2-1. Incidents that occurred during a normal process or facility shutdown are not included as the equipment is not in transition—they are in their stable, at rest state. Note that this figure was developed from the following timelines and is based on a continuous process:

- Normal operations (Figure 3.2)
- Process or facility shutdowns (Figure 4.3)
- Abnormal operations (Figure 6.3 and Figure 7.1)
- Emergency operations (Figure 8.2 and Figure 8.3)

It is important to recognize that there are several start-up-related steps that should be addressed depending on the type of preceding shut-down that occurred. Incidents have occurred when the operations group has attempted their restart without understanding why the process was shut-down in the first place or what "final" condition the equipment was in after the shut-down. The different shut-downs and the different responses and subsequent start-ups corresponding the them are summarized in Table A.2-1. Safe, incident-free restarts require the Operational Discipline (OD) of everyone to understand and address any issues, no matter whether it was an unexpected, unscheduled, or emergency shut-down. The effect of weak OD was discussed in Chapter 10, showing with the simplified risk equation that by increasing OD across all levels in an organization helps reduce its process safety risks (Equation 10.1). As noted earlier in this guideline, process and facility shutdowns have had start-up-related issues when equipment-related issues arose when preparing for the commissioning step, as well.

Further analysis of the incident results has been drafted at the time of publication of this book [108]. The paper addresses process safety culture and leadership and its impact on the facility's operational discipline to manage its start-ups and shut-downs effectively and safely. The conclusion (similar to the higher-level assessment in this guideline) is that the process safety risk for incidents is higher during start-ups and shut-downs *when the engineering controls have not been maintained and the administrative controls are not effective.*

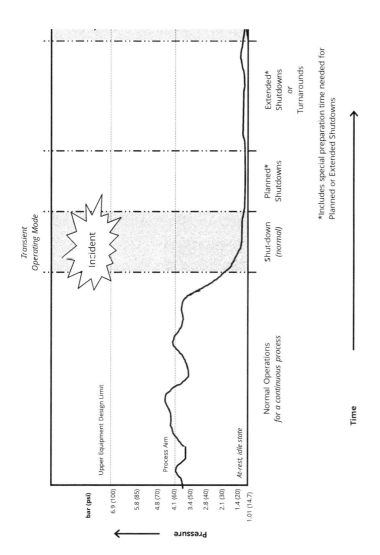

Figure A.2 1 Timeline of when incidents occurred during the transient operating mode.

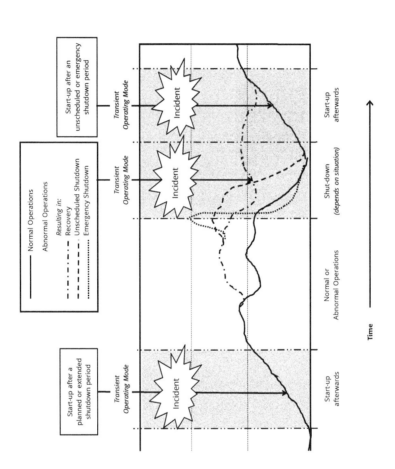

Figure A.2-1Timeline of when incidents occurred during the transient operating mode (continued).

Table A.2 1Phases for the transient operating modes.

Figure for Continuous Process	Facility Operation (Note)	Phases During Operations			Shutdown Note	Applicable Steps Before Resuming Normal Operation				
		Start-up	Recovery	Shut-down		Activate Emergency Response Team (ERT)	Investigate for Cause(s)	Address Issues	Address Special Start-up?	Startup
3.2	Normal Operation	Normal	n/a	Normal	Ends at a normal safe, idle, and at-rest state	n/a	n/a	n/a	n/a	Normal
6.3	Abnormal Operation	Normal	Recovery	n/a	n/a	n/a	Possible, if cause(s) identified	Possible, if issues identified	n/a	n/a
7.1	Abnormal Operation	Normal	n/a	Unscheduled	Ends at a normal safe, idle, and at-rest state	Site Specific	Yes	Yes, if issues identified	n/a	Normal
8.2	Abnormal Operation abruptly transitioning to the Emergency Operation	Normal	n/a	Emergency	Ends at a normal safe, idle, and at-rest state May activate the Emergency Response Team (ERT)	Site Specific	Yes	Yes, if issues identified	n/a	After emergency shutdown - normal
8.3	Abnormal Operation abruptly transitioning to the Emergency Operation	Normal	n/a	Emergency	Ends at a different but safe state May activate the Emergency Response Team (ERT)	Site Specific	Yes	Yes	Yes	After emergency shutdown, addresses different end state
A.2-1	Emergency Operation initiated by a loss event	Normal	n/a	Emergency	Activates the Emergency Response Team (ERT) May activate the Emergency Response Plan (ERP)	Yes	Yes	Yes	Yes, if different end state	After emergency shutdown, addresses different end state, if needed

The incidents compiled for this guideline are listed in Table A.2-2; the RBPS element weaknesses are noted in this table and are discussed in more detail in Chapter 10. Only a few of these incidents were discussed in detail in other chapters, with the references for the rest of the incidents noted as follows:

- Multi-incident references include
  - (CCPS 2007) [109]
  - (CCPS 2018) [67]
  - (CCPS 2019) [76]
  - (Marsh 2018) [36]
  - (Khan 2017) [110]
  - (Kletz 2009) [20]
  - (Sanders 2015) [2]
- CSB Website [111]
  - (CSB 2003a) Distillation Column Standby [73]
  - (CSB 2003b) DPC Enterprises [83]
  - (CSB 2005) Honeywell HF Release [84]
  - (CSB 2007) BP Texas City [47]
  - (CSB 2011a) Bayer CropScience [50]
  - (CSB 2011b) DuPont Belle Plant [112]
  - (CSB 2015a) Millard Refrigerated Sys. [18]
  - (CSB 2015b) DCRC [39]
  - (CSB 2015c) Chevron Richmond [61]
  - (CSB 2017) ExxonMobile Torrance Refinery [113]
  - (CSB 2018a) Husky Refinery [45]
  - (CSB 2018b) Arkema Crosby [78]
  - (CSB 2018c) Start-ups and Shut-downs [5]
  - (CSB 2018d) Kuraray America [114]
- CCPS Process Safety Beacon Website: [115]
  - (Beacon 2002) Vacuum Incidents [38]
  - (Beacon 2011) ConAgra and Kleen Energy [92]
  - (Beacon 2013a) Water Pump Explosions [116]
  - (Beacon 2013b) Pneumatic Testing [90]
  - (Beacon 2019) "Wrong" Instrument Reading [117]

- Other References in Table A.2-2
  - o (Behie 2008) Dolphin Energy [44]
  - o (Bloch 2016) Bhopal [95]
  - o (EPA 2018) Tosco Avon Refinery [118]
  - o (EPSC 2019) Flare System [119]
  - o (Fogler 2011) Monsanto [62]
  - o (Meshkati 2014) [77]
  - o (NFPA 2011) Hydrocracker Excursions [120]
  - o (UK HSE 1997) Texaco Pembroke [68]

Table A.2 2 Summary of the incidents during the transient operating mode.

Incident	See	Risk Based Process Safety Element / Year	Transient Operating Mode	Process Safety Culture (1)	Compliance with Standards (2)	Process Safety Competency (3)	Workforce Involvement (4)	Stakeholder Outreach (5)	Process Knowledge Management (6)	Hazard Identification and Risk Analysis (7)	Operating Procedures (8)	Safe Work Practices (9)	Asset Integrity and Reliability (10)	Contractor Management (11)	Training and Perform. Assurance (12)	Management of Change (13)	Operational Readiness (14)	Conduct of Operations (15)	Emergency Management (16)	Incident Investigation (17)	Measurement and Metrics (18)	Auditing (19)	Management Review and Contin. Improv. (20)
			53% | 47%						Pillar II		Pillar III									Pillar IV			
			35 | 31																				
Elements Identified as "week" (See Figure 10.3)				5%	3%	6%	1%	2%	9%	15%	10%	3%	12%	1%	3%	8%	2%	3%	8%	4%	1%	1%	2%
No. of Identified RBPS Causes		Year		16	12	20	3	6	30	52	35	10	40	5	12	27	8	11	29	15	4	3	7
Chapter 1 Introduction																							
Millard Refrig. Sys. (CSB 2015a)	C7.6.3-1																						
Chapter 4 - Table 1.1 Modes 3, 4 Planned Shutdowns																							
DCRC Flash Fire (CSB 2015b)	C4.7.1-1	2015	1						1	1	1					1							
(Shut-down) Not discussed in Chapter 4																							
Replacing Oil and Valves (Sanders 2015; p. 153)	(C4.7.1) (A.4-1)	Not Known	1						1	1	1					1							
Tank Cleaning (Sanders 2015; p. 155)	(C4.7.1) (A.4-1)	Not Known	1						1	1	1												
Vacuum Incidents, Cleaning (Beacon 2002)	(C4.7.1) (A.4-1)	Not Known	1					1	1				1										
Replacing Valves (Sanders 2015; p. 156)	(C4.7.1) (A.4-1)	Not Known	1						1	1	1												
Maintenance Replacement (Sanders 2015)	C4.7.2-1	Not Known	1							1	1		1			1	1	1					
(Start-up) Not discussed in Chapter 4																							
Repaired, Cold Furnace (Kletz 2009; p. 327)	(C4.7.2) (A.4-1)	Not Known	1						1		1		1		1	1							
Water Pump Explosions (Beacon 2013a)	(C4.7.2) (A.4-1)	Not Known	1						1		1		1										

Table A.2-2 Summary of the incidents during the transient operating mode (Continued)

Incident	Risk Based Process Safety Element / Year	Transient Operating Mode	Commit to Process Safety (Pillar I)	1 Process Safety Culture	2 Compliance with Standards	3 Process Safety Competency	4 Workforce Involvement	5 Stakeholder Outreach	Understand Haz. and Risks (Pillar II)	6 Process Knowledge Management	7 Hazard Identification and Risk Analysis	Manage Risk (Pillar III)	8 Operating Procedures	9 Safe Work Practices	10 Asset Integrity and Reliability	11 Contractor Management	12 Training and Perform. Assurance	13 Management of Change	14 Operational Readiness	15 Conduct of Operations	16 Emergency Management	Learn from Experience (Pillar IV)	17 Incident Investigation	18 Measurement and Metrics	19 Auditing	20 Management Review and Contin. Improv.
(column #)				1	2	3	4	5		6	7		8	9	10	11	12	13	14	15	16		17	18	19	20
No. of Identified RBPS Causes	Year	53% / 35	1	16	12	20	3	6		30	52		35	10	40	5	12	27	8	11	29		15	4	3	7
Elements Identified as "weak" (See Figure 10.3)		47% / 31		5%	3%	6%	1%	2%		9%	15%		10%	3%	12%	9%	3%	8%	2%	3%	8%		4%	1%	1%	2%
Chapter 5 – Table 1.1 Modes 5, 6 **Extended Shutdowns**																										
C5.6.1-1 — Dolphin Energy Ltd. (Behie 2008)	2009		1																							
C5.6.2.1 — Husky Superior Refinery (CSB 2018a)	2018	1									1				1											
C5.6.3-1 — BP Texas City (CSB 2007)	2005	1		1	1		1			1	1		1	1	1	1	1	1	1	1	1		1	1	1	1
(Start-up) Not discussed in Chapter 5																										
C5.6.3 (A.4-1) — Bayer CropScience (CSB 2011a)	2008	1		1					1	1	1		1	1	1			1	1	1	1	1				
C5.6.3 (A.4-1) — Steam Generation (Sanders 2015; p. 79)	1991	1																		1						
C5.6.3 (A.4-1) — Ethylene Plant Start-up (Kletz 2009; pp. 408-411)	Not Known	1			1	1							1		1			1	1		1		1	1	1	
C5.6.3 (A.4-1) — Start-up Afterwards (Kletz 2009; pp. 252)	Not Known	1													1											
Chapter 6 Recovery																										
C6.5-1 — Monsanto 1969 (Fogler 2011)	1969			1						1	1		1		1				1				1		1	1

Table A.2-2 Summary of the incidents during the transient operating mode (Continued)

Incident	Risk Based Process Safety Element / Year	Transient Operating Mode	Pillar I – Commit to Process Safety					Pillar II – Understand Haz. and Risks		Pillar III – Manage Risk									Pillar IV – Learn from Experience			
			1 Process Safety Culture	2 Compliance with Standards	3 Process Safety Competency	4 Workforce Involvement	5 Stakeholder Outreach	6 Process Knowledge Management	7 Hazard Identification and Risk Analysis	8 Operating Procedures	9 Safe Work Practices	10 Asset Integrity and Reliability	11 Contractor Management	12 Training and Perform., Assurance	13 Management of Change	14 Operational Readiness	15 Conduct of Operations	16 Emergency Management	17 Incident Investigation	18 Measurement and Metrics	19 Auditing	20 Management Review and Contin. Improv.
Elements identified as "weak" (See Figure 10.3)		47%	5%	3%	6%	1%	2%	9%	15%	10%	3%	12%	1%	3%	8%	2%	3%	8%	4%	1%	1%	2%
No. of Identified RBPS Causes	Year	31	16	12	20	3	6	30	52	35	10	40	5	12	27	8	11	29	15	4	3	7
Chapter 7 – Table 1.1 Modes 7, 8																						
Unscheduled Shutdowns																						
C7.6.1-1 Batch Vacuum Still Standby (Kletz 2009; pp. 287-288)	Not Known	1																				
C7.6.2-1 Distillation Column Standby (CSB 2003a)	2002	1																	1			
(Standby) Not discussed in Chapter 7																						
C7.6.2 (A.4-1) ExxonMobil Torrance (CSB 2017)	2015	1		1				1	1	1		1			1							
C7.6.3-1 Millard Refrig. Sys. (CSB 2015a)	2010	1			1			1	2	1	1	1		1								
C7.6.4.1-1 Fukushima Nucl. Power (CCPS 2019)	2011	1		1			1	1	1		1	1		1				1				
C7.6.4.2-1 Hurricane Georges (March 2018)	1998	1			1			1		1	1	1						1				
C7.6.4.2-2 Arkema Crosby (CSB 2018b)	2017	1						1	1		1							1				
(Start-up) Not discussed in Chapter 7																						
C7.6.4.2 (A.4-1) After lightning strike (Kletz 2009; pp. 549-552)	Not Known	1					1	1				1			1			1				
C7.6.4.2 (A.4-1) Texaco Pembroke (UK HSE 1997)	1994	1						1		1		1			1				1			
C7.6.4.2 (A.4-1) Esso Longford (Khan 2017; Chap 2)	1998	1		1	1			1		1		1		1			1	1	1	1		
C7.6.4.3-1 Hydrological Events	None Ident.																	1				
C7.6.4.4-1 Ice Storm (CCPS 2018)	Not Known	1						1														

Table A.2-2 Summary of the incidents during the transient operating mode (Continued)

Incident	Risk Based Process Safety Element / Year	Transient Operating Mode	Process Safety Culture (1)	Compliance with Standards (2)	Process Safety Competency (3)	Workforce Involvement (4)	Stakeholder Outreach (5)	Process Knowledge Management (6)	Hazard Identification and Risk Analysis (7)	Operating Procedures (8)	Safe Work Practices (9)	Asset Integrity and Reliability (10)	Contractor Management (11)	Training and Perform. Assurance (12)	Management of Change (13)	Operational Readiness (14)	Conduct of Operations (15)	Emergency Management (16)	Incident Investigation (17)	Measurement and Metrics (18)	Auditing (19)	Management Review and Contin. Improv. (20)
			Pillar I Commit to Process Safety					*Pillar II* Understand Haz. and Risks		*Pillar III* Manage Risk									*Pillar IV* Learn from Experience			
			1	2	3	4	5	6	7	8	9	10	11	12	13	14	15	16	17	18	19	20
Elements identified as "weak" (See Figure 10.3)		53% / 47%	5%	3%	6%	1%	2%	9%	15%	10%	3%	12%	1%	3%	8%	2%	3%	8%	4%	1%	1%	2%
No. of Identified RBPS causes	Year	35 / 31	16	12	20	3	6	30	52	35	10	44	5	12	27	8	11	29	15	4	3	7
Chapter 8 - Table 1.1 Modes 9, 10																						
Emergency Shutdowns																						
C8.6.1-1 Caused by emergency shut-down	None Ident.																					
C8.6.2-1 DPC Enterprises, L.P. (CSB 2003b)	2002	1																				
(Shut-down) Not discussed in Chapter 8																						
C8.6.2) (A.4-1) Chevron Richmond Refinery (CSB 2015c)	2012	1										8						1	1			
C8.6.3-1 Honeywell HF Release (CSB 2005)	2003	1				1			1		1							1				
Chapter 9 - Initial Startup (project)																						
Life Cycle Incidents Pneumatic Testing (Beacon 2013b)	2007	1									1					1						
C9.4.1-2 New Filters in Piping (Sanders 2015)	Not Known	1						1	1		1					1						
(Commissioning) Not discussed in Chap 9																						
C9.4.1 (A.4-1) Storage Tank (Kletz 2009; p. 463)	Not Known				1			1	1				1		1		1					
C9.4.1 (A.4-1) Redundant Pressure Vessel (Kletz 2009; p. 181)	Not Known	1								1	1		1			1			1			
C9.4.1 (A.4-1) Kuraray America (CSB 2018d)	2018	1													1							
C9.4.2-1 ConAgra and Kleen Energy (Beacon 2011)	2010	1							1	1	1						1					

Table A.2-2 Summary of the incidents during the transient operating mode (Continued)

Column key (Risk Based Process Safety Elements):

Pillar I – Commit to Process Safety: 1 Process Safety Culture · 2 Compliance With Standards · 3 Process Safety Competency · 4 Workforce Involvement · 5 Stakeholder Outreach

Pillar II – Understand Haz. and Risks: 6 Process Knowledge Management · 7 Hazard Identification and Risk Analysis

Pillar III – Manage Risk: 8 Operating Procedures · 9 Safe Work Practices · 10 Asset Integrity and Reliability · 11 Contractor Management · 12 Training and Perform. Assurance · 13 Management of Change · 14 Operational Readiness · 15 Conduct of Operations · 16 Emergency Management

Pillar IV – Learn from Experience: 17 Incident Investigation · 18 Measurement and Metrics · 19 Auditing · 20 Management Review and Contin. Improv.

Incident	Year	Transient Operating Mode	1	2	3	4	5	6	7	8	9	10	11	12	13	14	15	16	17	18	19	20
Elements identified as "week" (See Figure 10.3)		53% / 47%	5%	3%	6%	1%	2%	9%	15%	10%	3%	12%	1%	3%	8%	2%	3%	8%	4%	1%	1%	2%
No. of Identified RBPS Causes	35	31	16	12	20	3	6	30	52	35	10	40	5	12	27	8	11	29	15	4	3	7
(Start-up) Not discussed in Chapter 9																						
Condensing Steam (C9.4.2) (A.4-1) (Sanders 2015; p. 75)	Not Known	1													1							
Pneumatic Testing (C9.4.2) (A.4-1) (Kletz 2009; p. 370)	Not Known	1												1								
Pipework Debris (C9.4.2) (A.4-1) (Kletz 2009; p. 370)	Not Known	1										1										
Chapter 9 – Mothballing Life Cycle Incidents																						
Mothballed Vessel (C9.7-1) (Kletz 2009; p. 455)	Not Known	1														1						
Deconm Mothball (C9.7-2)	None Ident.																					
Chapter 9 – Decommissioning Life Cycle Incidents																						
Redundant Piping Removal (C9.9.1-2) (Kletz 2009; p. 367-368)	Not Known	1							1	1	1	1	1	1	1		1	1	1	1	1	1
Bhopal (C9.9.2-1) (Bloch 2016)	1984	1		1		1		1	1	1	1	1	1	1	1		1	1	1	1	1	1
Decommissioning (Not Discussed)																						
DuPont Belle Plant (C9.9.1) (A.4-1) (CSB 2011b)	2010	1	1					1			1							1	1			

Table A.2-2 Summary of the incidents during the transient operating mode (Continued)

Incident	Risk Based Process Safety Element	Transient Operating Mode	1 Process Safety Culture	2 Compliance with Standards	3 Process Safety Competency	4 Workforce Involvement	5 Stakeholder Outreach	6 Process Knowledge Management	7 Hazard Identification and Risk Analysis	8 Operating Procedures	9 Safe Work Practices	10 Asset Integrity and Reliability	11 Contractor Management	12 Training and Perform. Assurance	13 Management of Change	14 Operational Readiness	15 Conduct of Operations	16 Emergency Management	17 Incident Investigation	18 Measurement and Metrics	19 Auditing	20 Management Review and Contin. Improv.
Elements Identified as "weak" (See Figure 10.3)	53%	47%	5%	3%	6%	1%	2%	9%	15%	10%	3%	12%	1%	3%	8%	2%	3%	8%	4%	1%	1%	2%
No. of Identified RBPS Causes (Year)	35	31	16	12	20	3	6	30	52	35	10	40	5	11	27	8	11	29	15	4	3	7

Start-up or Shut-down transient operating mode incidents from: CCPS 2019 (More Incidents the Define Process Safety)

Incident	Transient Operating Mode	1	2	3	4	5	6	7	8	9	10	11	12	13	14	15	16	17	18	19	20
T2 Laboratories, Inc. Runaway reaction	1	1	1	1				1						1			1				
Millard Refriger. Serv. Ammonia	1	1																			
Hoechst Griesheim Runaway reaction	1			1			1						3								
Arco Channelview Explosion	1	1	1	1																	
Port Neal, USA AN Explosion	1	1																1			
Hickson & Welch Jet Fire	1							1	1												
Chevron Richmond Refinery fire	1							1	1		1										
Buncefield depot Storage Tank	1								1							1					
Celanese Pampa Explosion									1		1			1		1	1				
Hayes Lemmerz Dust explosion	1	1						1	1					1			1				
Macondo Well Deepwater Horizon	1		1					1	1		1			1		1	1				
DuPont LaPorte Methyl Mercaptan	1						1		1							1	1	1			
DPC Enterprises Chlorine	1								1					1		1	1				
Gaylord Chemical Nitrogen Tetroxide	1	1										1		1		1	1	1			
Fukushima Daiichi Nuclear Plant	1										1			1			1				

Table A.2-2 Summary of the incidents during the transient operating mode (Continued)

Incident	Transient Operating Mode 53%	Transient Operating Mode 47%	1 Process Safety Culture	2 Compliance with Standards	3 Process Safety Competency	4 Workforce Involvement	5 Stakeholder Outreach	6 Process Knowledge Management	7 Hazard Identification and Risk Analysis	8 Operating Procedures	9 Safe Work Practices	10 Asset Integrity and Reliability	11 Contractor Management	12 Training and Perform. Assurance	13 Management of Change	14 Operational Readiness	15 Conduct of Operations	16 Emergency Management	17 Incident Investigation	18 Measurement and Metrics	19 Auditing	20 Management Review and Contin. Improv.
			Pillar I Commit to Process Safety					Pillar II Understand Haz. and Risks		Pillar III Manage Risk									Pillar IV Learn from Experience			
Risk Based Process Safety Safety Element	1		1	2	3	4	5	6	7	8	9	10	11	12	13	14	15	16	17	18	19	20
Year	35	31																				
No. of Identified RBPS Causes (Elements Identified as "weak" (See Figure 10.3)) — %			5%	3%	6%	1%	2%	9%	15%	10%	3%	12%	1%	3%	8%	2%	3%	8%	4%	1%	1%	2%
No. of Identified RBPS Causes — count			16	12	20	3	6	30	52	35	10	40	5	12	27	8	11	29	15	4	3	7
Start-up or Shut-down transient mode incidents from: CCPS 2009 (Incidents the Define Process Safety)																						
Three Mile Island Nuclear Plant	1							1	1			1						1	1			
Chernobyl Nuclear Plant, Russia	1							1	1			1			1			1				1
PEMEX Terminal Mexico City	1								1			1						1				
Tupras Refinery Turkey Earthquake	1							1	1			1						1				
Texaco Oil Refinery Milford Haven	1							1	1			1			1			1				
Elf Refinery BLEVE Feyzin, France	1								1	1		1			1			1	1			
BP Grangemouth UK	1							1	1	1		1			1			1				1
Esso Longford Australia		1							1	1		1			1			1				1
Flixborough UK		1							1			1			1							
Avon Refinery Tosco CA		1						1	1	1		1			1			1	1			1
BP Texas City		1							1	1		1			1			1	1			1

A.3 Managing the unexpected during transient operating modes

Although there were no references to the "expecting the unexpected" concept in the transient operating modes incidents, the concept is not new [121]. Based on the significant process safety incidents that have occurred during all operating modes, there is much room for applying the concept within each mode. Since there should be an understanding what types of unexpected situations that can occur, this section provides more detailed guidance on how best to anticipate unplanned but "expected" shut-down situations for:

- Loss of utilities (Section A.3.1)
- Engineering control issues (Section A.3.2)
- Administrative control issues (Section A.3.3)

This section concludes with a brief discussion on the dangers of "normalizing the deviance" and maintaining a sense of vulnerability when managing the risks of hazardous processes (Section A.3.4).

A.3.1   Anticipating and addressing unplanned loss of utilities

As was noted in Chapter 7, a hazards analysis team can use a checklist for evaluating the loss of utilities, helping the team anticipate issues and establishing procedures for a safe shut-down and a safe restart. At some point during normal operations, for example, there could be a sudden, unexpected power outage that could shut down the entire facility. For this reason, facilities have developed safeguards to ensure that there is a reliable backup emergency electrical supply that can be used to safely shutdown the affected equipment. A useful checklist for other potential utility losses that could jeopardize the engineering controls required to manage the processes safely is provided in Table A.3-1.

Table A.3 1 Loss of utilities checklist.

	Utility	Equipment Effected by Loss
1	Electricity	Effect on PLC, interlocks, pumps, etc.; power surges
2	Electrical Backup Systems	Generators, Uninterruptible Power Supply (UPS)
3	Air	Instrument or process air (dry; wet; contaminated); high pressure air; pneumatic conveying systems
4	Inerting Gases	Inerting or purging atmospheres; nitrogen; argon; contaminated
5	Fuels (gas)	Natural gas (including Liquid Petroleum Gas, LPG), hydrogen
6	Fuels ("hot oils")	Dowtherm, Therminol
7	Steam	Distribution, generation (boilers), hot condensate
8	Cooling systems	Cooling water, refrigeration, air conditioning, glycol
9	Water	Cooling (see above), process, hot water (vs. steam)
10	Wastewater Treatment	Process, sanitary, sewer (see Biological cross reference)
11	Vacuum	(see Pressure)
12	Incinerator	Thermal oxidizers
13	Emergency Response Systems	Halon systems, fire water supply, sprinkler water supply

A.3.2 Anticipating and addressing unplanned engineering control issues

Chapter 6 introduced an example matrix that listed some of the guidewords used in a Hazards and Operability (HAZOP) study for a continuous process (Table 6.1). A more comprehensive guideword matrix is presented in Table A.3-2 for both continuous and batch operations. Although these guidewords focus on deviations from the normal operating parameters, the guidewords can be used to anticipate and help establish the safe parameters associated with the shut-down and start-up procedures, as well.

The HAZOP Team will investigate the equipment design, as well, based on the scenarios being reviewed. The equipment should be fail-safe, with features that automatically counteract the effect of an anticipated deviation such as a power loss. A system is fail-safe if the failure of a component, signal, or utility, initiates action that returns the system to a safe condition [34]. Depending on the scenario and process application, a fail-safe valve may need to close (or remain closed), open (or remain open), or remain unchanged in its current operating position, whether fully closed, fully open, or anywhere in between. Thus, for unexpected shut-downs of the equipment, it should fail-safe.

In addition, hazards analysis teams should know the history of issues that have occurred as they develop potential scenarios. Potential fire, explosion, or toxic release issues during transient operating modes with start-ups and shut-downs include the following items (Adapted from [8, p. 7]):

1. Fires burning or resulting in an explosion when fuel mixes with oxygen in the presence of an ignition source.
2. Explosions that damage nearby equipment causing additional releases of other flammable materials that may then ignite and burn (the domino effect).

Table A.3 2 Hazard and Operability (HAZOP) Study guideword matrix.

| Guideword | Design Parameter ("Design Conditions") | | | | | | | Batch Operations | |
| | Continuous or Batch Operations | | | | | | | | |
	Flow	Temperature	Pressure	Level (Interface)	Composition	State	Reaction	Time	Sequence
More	High Flow	High Temperature	High Pressure	High Level	High Concentration	Additional Phase	High Rate	Too Long	Step Too Late
Less	Low Flow	Low Temperature	Low Pressure	Low Level	Low Concentration	Loss of Phase	Low Rate	Too Short	Step Too Early
None	No Flow	Cryogenic	Vacuum Pressure	No Level			No Reaction	Not Started	Step Left Out
Part Of					Wrong Concentration		Incomplete Reaction		Part of Step Left Out
As Well As			High/Low P Interface	Liq/Liq Interface	Contaminants / Impurities	Contaminants / Impurities	Side Reaction		Extra Action in Step
	Reverse				Wrong Material	Wrong Material	Reverse Reaction		Step Backwards
Other Than	Misdirected Flow					Change of State	Wrong Reaction	Wrong Time	Wrong Action Taken

Table A.3 2 Hazard and Operability (HAZOP) Study guideword matrix (Continued).

	Leak/Rupture (Heat exchanger Tube) Leak/Rupture (Vessel, Tank) Mixing Purging/Inerting Sampling, pH Static Charge Current Voltage	
Other Parameters include:	Node-specific Startup/Shutdown hazards Emergency operations Special maintenance hazards (mechanical and/or electrical) Corrosion/Erosion hazards Special safety concerns (in addition to normal PPE)	
	Area-specific (across all nodes) Global hazards (e.g., Corrosion under insulation, Grounding, etc.) Human Factors Facility Siting	

3. Destructive pressures that compromise or destroy piping and equipment, including:
   a. Large pressure surges when water is inadvertently mixed with hot oil;
   b. Large pressure surges when condensed materials in cooled equipment is heated and expands rapidly;
   c. Excessive internal pressures when vessel or piping contents are heated without room for expansion (e.g., a blocked-in heat exchanger);
   d. Vacuums in blocked-in vessels and equipment when hot materials cool and condense (such as steam used in cleaning *during* the shut-down preparations *for* a shutdown).

When water is not drained or its flow is lost, it may freeze, with the ice expanding and damaging the piping (i.e., in "dead legs") or equipment. Unfortunately, the freeze-related damage is usually detected after the ambient or process material's temperature rises, the ice thaws, and there is a loss of containment of the hazardous materials.

Thus, transition times may have unusual situations with unusual scenarios that needed to be addressed by the PHA Team. A series of transient operating mode guidewords used for the HAZOP performed for a safe and eco-friendly biomass gasification plant is provided elsewhere [122, p. 9/33]. These guidewords are as follows with an example deviation (an "abnormal operation") depicted using the following guidewords [122, pp. Appendix B, 30ff]:

- Transient operation—start-up
- Transient operation—shut-down
- Transient operation—increase plant power load
- Transient operation—emergency shut-down

As discussed in Chapter 10, many assumptions should be made by a Process Hazard Analysis (PHA) Team when performing a HAZOP study. It is not the purpose of PHA Team to evaluate the effectiveness of the other RBPS elements nor does the PHA Team have the time to determine which of the other elements are weak, since weaknesses in the other RBPS elements should be reviewed through a separate, structured auditing program (i.e., Element 19, Figure 10.1; also see [40]). Recognizing that if any of these assumptions are incorrect (there is a failure in a separate RBPS element), incidents will occur during *any operating mode*—normal, abnormal, emergency, and transient—no matter how many scenarios were evaluated by the PHA Team.

The fundamental hazards analysis team assumptions should be understood by the PHA Team *and by incident investigation teams*. Due to the interactions between RBPS elements, simply listing an inadequate Hazards Identification and Risk Analysis element may not adequately capture what actually lead to the incident, especially if there is a combination of other causes from other RBPS elements. The PHA Team should not be expected to anticipate combinations of all the causes using the HAZOP approach. An example of an investigation team assuming too much responsibility for the PHA Team was discussed elsewhere [123, p. Appendix D].

Since it is simply not possible to anticipate and establish shut-down and start-up procedure for every deviation combination, it is essential that the assumptions established for the PHA Team are documented and clearly communicated to everyone managing the process safety programs or elements. A comprehensive list of assumptions that can be made by the PHA Team during their HAZOP studies is provided in Table A.3-3. These general assumptions should be assessed during incident investigations, as not every scenario that could foreseeably appear can be anticipated, especially when more than one engineering or administrative control fails.

Table A.3 3 List of Process Hazard Analysis (PHA) team HAZOP assumptions

1	Assumptions during the HAZOP scenario development:	Often considered when evaluating transient operating modes
1.1	The scenarios are based on one component deviating at a time (e.g., a normally open valve is closed or closes unexpectedly).	Abnormal operations
1.2	Double jeopardy scenarios are not actively analyzed but are considered if the team believes the scenario is credible (e.g., opening/closing multiple valves during a lineup, or pump trips and the check valve does not hold, resulting in reverse flow).	Abnormal operations
1.3	An administrative control program is in place and adequate for ensuring that valves in pressure protection paths remain open (exceptions are noted in the PHA).	Administrative controls
1.4	All bypasses around control valves are normally closed (exceptions are noted in the PHA).	Normal operations
1.5	Misdirected flow will not be considered if the line is blinded either from the process or from being open to the environment.	Abnormal operations
1.6	Guards and barriers are in place to reduce the likelihood of external forces such as maintenance activities or vehicular traffic impacting process piping or equipment.	
1.7	Vehicular traffic is limited through the plant. Piping is routed and equipment is located such that the potential for vehicle impact to piping and equipment is minimized.	
1.8	External forces, such as tidal waves, hurricanes, typhoons, sandstorms, lighting, tornadoes, acts of sabotage, etc. will not typically be considered.	Extreme weather
1.9	For each scenario depicted in the HAZOP worksheets, the following expected ("global") safeguards are not specifically listed as safeguards: Operator training, qualifications, and rounds; Maintenance training and qualifications; and Fire protection and emergency response.	

Table A.3 List of Process Hazard Analysis (PHA) team HAZOP assumptions (continued)

2	Assumptions for Element 6, Process Knowledge Management, include:	Often considered when evaluating transient operating modes
2.1	The equipment under this review is fit for purpose - designed, fabricated, installed, operated, and maintained per "Recognized and Generally Accepted Good Engineering Practices (RAGAGEP)" or "Best Demonstrated Available Technology (BDAT)."	Unexpected equipment failure
2.2	The piping and instrumentation diagrams (P&IDs) for this review reflect the existing process equipment, instrumentation, and controls, and if applicable, reflect all modifications to any new or removed equipment since the last review.	Change management
2.3	The relief system design documentation includes verification of its sizing and applicability (e.g., materials of construction compatibility with process materials and types of flow, such as two-phase flow). Unless noted in the HAZOP worksheet, the relief system vents to a safe location when activated.	Unexpected equipment failure
2.4	The electrical classification for the equipment is correct for the hazardous materials being processed in the unit.	
2.5	Fire protection and mitigation equipment is installed, adequately sized, functional, and tested as necessary to ensure functionality and reliability.	Unexpected equipment failure
3	Assumptions for Element 8, Operating Procedures, include:	
3.1	The operating procedures describe each operating phase (start-up, normal shut-down, temporary, emergency shutdown, and start-up after emergency shut-down, if applicable).	Transient operations
3.2	The operators are adequately trained to perform their work duties.	All operations
3.3	The process is running under normal conditions and as designed before the deviation occurs	Normal operations

Table A.3 List of Process Hazard Analysis (PHA) team HAZOP assumptions (continued)

		Often considered when evaluating transient operating modes
4	**Assumptions for Element 10, Asset Integrity and Reliability, include:**	
4.1	The maintenance procedures are written, up-to-date, understood, and followed.	
4.2	The mechanics and electricians are trained to perform their work duties.	
4.3	The process and safety alarms are tested as necessary to ensure their reliability and functionality.	Unexpected equipment failure
4.4	The facility has a maintenance Inspection, Testing, and Preventive Maintenance program (ITPM).	Unexpected equipment failure
5	**Assumptions for Element 12, Training and Performance Assurance, include:**	
5.1	All personnel, including operators and maintenance personnel, are trained for, understand their specific roles, and follow the facility's safe work permits, including those specifically written for hot work, vessel entry (confined space), and electrical isolation (e.g., Lock-Tag-Try or Lock-Out-Tag-Out).	Unexpected personnel or equipment failure

Table A.3 List of Process Hazard Analysis (PHA) team HAZOP assumptions (continued)

		Often considered when evaluating transient operating modes
6	**Assumptions for Element 15, Conduct of Operations (Operational Discipline`, include:**	
6.1	All personnel in the hazardous areas, operators, maintenance, contractors, and visitors, are wearing the area's Personal Protective Equipment (PPE), and are trained on how to properly inspect, don, and use the PPE.	Unexpected personnel failure
7	**Assumptions for Element 16, Emergency Management, include:**	
7.1	Emergency response plans are written and communicated to all personnel—employees as well as contractors.	Unexpected personnel failure
7.2	Evacuation signals are known and evacuation routes established.	Unexpected personnel failure
7.3	Emergency response equipment is inspected and maintained.	Unexpected equipment failure
7.4	Emergency responders are adequately trained to perform their work.	Unexpected personnel failure

A.3.3   Anticipating and addressing unplanned administrative control issues

Although an abnormal situation may cause an abnormal operation (see discussion in Chapter 6), the unexpected situation can lead to creative, well-intentioned workarounds or "one minute changes" by those performing the work, as well (see discussion in Chapter 1). These quick changes lead to increased risks to all parties involved and then—unfortunately—a serious incident which causes harm to people, the environment, and property. Thus, the unexpected situation should trigger everyone's "sense of vulnerability," as well [121].

Many of the incidents reviewed for this guideline, especially during start-ups, occurred during unexpected situations, when creative changes were implemented without a proper hazards identification and risk assessment. No matter how much detail is developed for the start-up procedure, there may be something that does not follow the pre-plan. Note that a pre-plan cannot anticipate all potential situations and should only be used as a guide during the emergency. During the emergency, unanticipated and never-happened-before situations will occur [81]. An effective project start-up that addresses unexpected issues could follow the guidance listed in Table A.3-4. Additional guidance for effectively managing unexpected situations to help prevent process safety incidents is provided in the literature [2, pp. 415-426].

A.3.4   Normalization of the deviance issues

An insidious behavior during normal operations occurs when a facility "normalizes the deviance" and accepts the deviation as normal operations [124]. Normalization leads to complacency and, ultimately, to the loss of a sense of vulnerability, as well. It is important to recognize that some incidents occurred during abnormal situations when personnel:

1) did not acknowledge that the conditions were not normal or as planned;

2) did not adequately address modified or non-routine activities; or, most often during normal operations,

3) ignored abnormal conditions altogether, sometimes for days or even months before the incident occurred [20, pp. 82-84] [84, pp. 85-87].

Given that some of these unexpected conditions provided ample time for a proper response, these abnormal conditions could have been recognized early and properly responded to proactively, helping prevent the loss event. Recognizing these early warning signals has been discussed in more detail elsewhere [125].

Table A.3-4 Steps for effectively managing unexpected situations during start-up

1	Plan the start-up (Establish a pre-plan – a start-up procedure)
2	Communicate the start-up pre-plan to all groups and teams involved
3	Execute the start-up pre-plan
4	**Expect the unexpected** (Recognize that pre-plans cannot address all start-up situations)
4a	If no unexpected issues arise, go to Step 7
4b	If unexpected situations arise, go to Step 5
5	Understand what is happening (Identify what is different than what was expected through the pre-plan)
6	Address unexpected issue safely by
6a	Identifying and addressing the unexpected issues
6b	Making changes to pre-planned start-up procedure
6c	Communicating the updated start-up plan to all groups and teams involved
6d	Resuming execution of the start-up with the updated start-up plan
6e	Go to Step 4
7	Debrief all team members once the process is running safely (When process is operating within its normal operating conditions)

References

[1] K. Davis, "The Safety Matrix: People Applying Technology to Yield Safe Chemical Plants and Products," in *Advances in Chemical Engineering, Vol. 14*, San Diego, CA USA, Academic Press, Inc., 1988, pp. 261-318.

[2] R. E. Sanders, Chemical Process Safety: Learning from Case Histories, Fourth Edition, Boston, USA: Elsevier, Butterworth-Heinemann, 2015.

[3] I. M. Duguid, "Analysis of Past Incidents in the Oil, Chemical and Petrochemical Industries," *IChemE Loss Prevention Bulletin 142*, no. 142, pp. 3-6, 1998.

[4] S. W. Ostrowski and K. Keim, "Tame Your Transient Operations: Use a special method to identify and address potential hazards," *Chemical Processing*, pp. 1-5, 23 June 2010.

[5] CSB, "CSB Investigations of Incidents During Startups and Shutdowns," US Chemical Safety and Hazard Investigation Board, Washington, D.C. USA, 2018.

[6] P. Okoh and S. Haugen, "Maintenance-related major accidents: Classification of causes and case study," *Journal of Loss Prevention in the Process Industries*, vol. 26, pp. 1060-1070, 2013.

[7] K. Kidam and M. Hurme, "Origin of equipment design and operation errors," *Journal of Loss Prevention in the Process Industries*, vol. 25, pp. 937-949, 2012.

[8] BP, "Safe Ups and Downs for Process Units (BP Process Safety Series)," IChemE, Rugby, UK, 2006.

[9] R. Jarvis and A. Goddard, "An Analysis of Common Causes of Major Losses in the Onshore Oil, Gas & Petrochemical Industries," Lloyd's Market Association, London, UK, 2016.

[10] B. Karthikeyan, "Moving Process Safety into the Board Room," *Chemical Engineering Progress,* no. September, pp. 43-46, 2015.

[11] OECD, "Corporate Governance for Process Safety: OECD Guidance for Senior Leaders in High Hazard Industries," OECD, Paris, France, 2012.

[12] T. Moore, "The failure of corporate management to equate process safety with production," *IChemE Loss Prevention Bulletin 257,* no. October, pp. 19-22, 2017.

[13] CCPS, Process Safety Leadership from the Boardroom to the Frontline, Hoboken, NJ USA: John Wiley & Sons, 2019.

[14] CCPS, Guidelines for Risk Based Process Safety (RBPS), New York, NY: AIChE, Wiley & Sons, 2007.

[15] CCPS, Guidelines for Hazard Evaluation Procedures, 3rd Edition, Hoboken , NJ: John Wiley & Sons, 2008.

[16] J. D. Kelly, "Formulating Production Planning Models," *Chemical Engineering Progress,* no. January, pp. 43-50, 2004.

[17] P. T. Otis and D. Hampson, "Improve Production Scheduling to Increase Energy Efficiency," *Chemical Engineering Progress,* no. March, pp. 45-51, 2017.

[18] CSB, "Key Lessons for Preventing Hydraulic Shock in Industrial Refrigeration Systems: Anhydrous Ammonia Release at Millard Refrigerated Services, Inc., Report 2010-13-A-AL," US Chemical Safety and Hazard Investigation Board, Washington, D.C. USA, 2015.

[19] B. Seggerman, "Ensuring Process Safety in Batch Tolling," *Chemical Engineering Progress,* no. December, pp. 34-40, 2017.

[20] T. A. Kletz, What Went Wrong? Case Histories and Process Plant Disasters and How They Could Have Been Avoided, Fifth Edition, Boston, USA: Elsevier, Gulf Publishing, 2009.

[21]   J. A. Klein and B. K. Vaughen, Process Safety: Key Concepts and Practical Approaches, Boca Raton, FL USA: CRCPress, 2017.

[22]   CCPS, Dealing with Aging Process Facilities And Infrastructure, Hoboken, NJ USA: John Wiley & Sons, 2018.

[23]   CCPS, Guidelines for Asset Integrity Management, Hoboken, NJ USA: John Wiley & Sons, 2017.

[24]   NDEP, "IV. Standard Operating Procedures (SOP) Data Form,Revision 3," Nevada Division of Environmental Protection (NDEP), Chemical Accident Prevention Program, Las Vegas, NV, USA, 2009.

[25]   CSB, "Catastrophic Rupture of Heat Exchanger, Report 2010-08-I-WA," US Chemical Safety and Hazard Investigation Board, Washington, D.C. USA, 2014.

[26]   CSB, "Investigation Report: Sterigenics (2004-11-I-CA)," US Chemical Safety and Hazards Investigation Board, Washinton, DC USA, 2006.

[27]   J. I. Gustin, "How the study of accident case histories can prevent runaway reaction accidents from recurring," *Trans IChemE,* Vols. 80, Part B, no. January, pp. 16-24, 2002.

[28]   N. Ramzan, S. Naveed, M. Rizwan and W. Witt, "Root Cause Analysis of Primary Reformer Catastrophic Failure: A Case Study," *Process Safety Progress,* vol. 30, no. 1, pp. 62-65, 2011.

[29]   CCPS, Guidelines for Investigating Process Safety Incidents, Hoboken, N. J.: John Wiley & Sons, 2019.

[30]   CCPS Beacon, "Interlocked for a Reason," Center for Chemical Process Safety, New York, NY USA, June 2003.

[31]   CCPS, Guidelines for Integrating Process Safety into Engineering Projects, Hoboken, NJ USA: John Wiley & Sons, 2019.

[32] MKOPSC, "Process Safety Taxonomy/Definitions," Mary Kay O'Connor Process Safety Center, wikips.tamu.edu, College Station, TX USA, 2019.

[33] CCPS, Guidelines for the Management of Change for Process Safety, Hoboken, NJ USA: John Wiley & Sons, 2008.

[34] CCPS, "CCPS Process Safety Glossary," Center for Chemical Process Safety, 2019. [Online]. Available: www.aiche.org/ccps.

[35] PMI, A Guide to the Project Management Body of Knowledge, Newton Square, PA: Project Management Institute, 2013.

[36] Marsh, "The 100 Largest Losses, 1978–2017, Large Property Damage Losses in the Hydrocarbon Industry, 25th Edition," Marsh & McLennan Companies, London, UK, 2018.

[37] CCPS, "The Safe Work Practices (SWP) Tool," Center for Chemical Process Safety, 2019. [Online]. Available: www.aiche.org/ccps.

[38] CCPS Beacon, "Vacuum is a powerful force!," Center for Chemical Process Safety, New York, NY USA, February 2002.

[39] CSB, "Key Lessons for Preventing Incidents When Preparing Process Equipment for Maintenance, Safety Bulletin No. 2015-01-I-DE," US Chemical Safety and Hazard Investigation Board, Washington, D.C. USA, 2017.

[40] CCPS, Guidelines for Auditing Process Safety Management Systems, Hoboken, NJ USA: John Wiley & Sons, 2011.

[41] CSB, Comments on the State of Washington Department of Labor and Industries Division of Occupational Health Proposed Rule, Washington, DC USA: U.S. Chemical Safety and Hazard Investigation Board, May 11, 2018.

[42] J. Perry, A. Hanna, S. Fernandez and G. Stevens, "Horses for Courses: Matching the Approach to Hazard Analysis With Project Schedule and

Design Defininition," in *Hazards XXII, Symposium Series No. 156, ICheme*, Rugby, UK, 2011.

[43] MTI, "Guidelines for Mothballing of Process Plants," Materials Technology Institute (MTI) of the Chemical Process Industries, Inc., Saint Louis, MO USA, 1989.

[44] S. Behie, L. Lackzo, J. Murtaza, S. Lied, N. Cooper, N. Aulmann and A. Al Rahbi, "Effective Management of the HSE&S Aspects of a Major Plant Shutdown," in *Society of Petroleum Engineers, 2010 SPE Middle East Health, Safety, Security, anjd Environmental Conference*, Manama, Bahrain, 2010.

[45] CSB, "Factual Investigative Update, Husky Superior Refinery Explosion and Fire," US Chemical Safety and Hazard Investigation Board, Washington, D.C. USA, 2018.

[46] J. A. Baker, F. L. Bowman, G. Erwin, S. Gorton, D. Hendershot, N. Leveson, S. Priest, I. Rosenthal, P. Tebo, L. D. Wilson and D. A. Wiegmann, "The Report of the BP US Refineries Independent Safety Review Panel," www.bp.com/bakerpanelreport, [Online], 2007.

[47] CSB, "Refinery Explosion and Fire, Report No. 2005-04-I-TX," US Chemical Safety and Hazard Investigation Board, Washington, D.C. USA, 2007.

[48] B. K. Vaughen and J. A. Klein, "What you don't manage will leak; A tribute to Trevor Kletz," *Process Safety and Environmental Protection*, vol. 90, no. 5, pp. 411-418, 2012.

[49] B. K. Vaughen, J. A. Klein and J. W. Champion, "Our Process Safety Journey Continues: Operational Discipline Today," *Process Safety Progress*, vol. 37, no. 4, pp. 478-492, 2018.

[50] CSB, "Pesticide Chemical Runaway Reaction, Pressure Vessel Explosion, Report No. 2008-08-I-WV," US Chemical Safety and Hazards Investigation Board, Washington, D.C. USA, 2011.

[51] M. Naderpour, S. Nazir and J. Lu, "The role of situation awareness in accidents of large-scale technological systems," *Process Safety and Environmental Protection,* vol. 97, pp. 13-24, 2015.

[52] CCPS, "The Risk Analysis Screening Tool (RAST)," Center for Chemical Process Safety, 2019. [Online]. Available: www.aiche.org/ccps.

[53] W. Li, Y. Sun, Q. Cao, M. He and Y. Cui, "A proactive process risk assessment approach based on job hazards analysis and resilient engineering," *Journal of Loss Prevention in the Process Industries,* vol. 50, pp. 54-62, 2019.

[54] F. Ibrahm and J. Lebowitz, "Essentials of Job Hazard Analysis," *Chemical Engineering Progress,* no. November, pp. 52-56, 2018.

[55] G. Song, Khan, F. and Yang, M., "Probabilistic Assessment of Integrated Safety and Security Related Abnormal Events: A Case for Chemical Plants," *Safety Science,* vol. 113, pp. 115-125, 2019.

[56] F. Lees, Lees' Loss Prevention in the Process Industries, 3rd Ed., M. S. Mannan, Ed., Burlington, MA USA: Elsevier Butterworth-Heineman, 2005.

[57] DHS, "CSAT Security Vulnerability Assessment (SVA) and Site Security Plan (SSP)," US Department of Homeland Security, 2019. [Online]. Available: www.dhs.gov/cisa/csat-sva-and-ssp.

[58] DHS, "Chemical Sector Cybersecurity Framework Implementation Guide," US Department of Homeland Security (DHS), www.us-cert.gov, Washington, D.C., 2015.

[59] C. L. Smith, "Process Control for the Process Industries, Part 1: Dynamic Characteristics," *Chemical Engineering Progress,* no. March, pp. 33-38, 2017.

References

243

[60] ASM Consortium, "Abnormal Situation Management," The Abnormal Situation Management Consortium, 2019. [Online]. Available: www.asmconsortium.org.

[61] CSB, "Chevron Richmond Refinery Pipe Rupture and Fire, Report 2012-03-I-CA," US Chemical Hazards and Safety Review Board, Washington, D.C. USA, 2015.

[62] H. S. Fogler, Essentials of Chemical Reaction Engineering, 1st Edition, Upper Saddle River, NJ USA: Prentice Hall, 2011.

[63] G. C. Vincent, "Rupture of a Nitroanaline Reactor," *Chemical Engineering Progress,* vol. 67, no. 6, pp. 51-57, 1971.

[64] USDA, "Help for Dealing with Plant Emergencies," US Department of Agriculture (USDA), Office of Outreach, Employee Education, and Training, Washington, DC USA, 2011.

[65] F. F. J. Liserio and P. W. Mahan, "Manage the Risks of Severe Wind and Flood Events," *Chemical Engineering Progress,* vol. 115, no. 4, pp. 42-49, 2019.

[66] CCPS Monograph, "Assessment of and planning for natural hazards," Center for Chemical Process Safety, New York, NY USA, 2019.

[67] CCPS, Guidelines for Siting and Layout of Facilities, Hoboken, NJ USA: John Wiley & Sonse, 2018.

[68] HSE, "The explosion and fires at the Texaco Refinery, Milford Haven (Pembroke), 24 July 1994," UK Health and Safety Executive, Norwich, UK, 1997.

[69] R. E. Sanders, "Hurrican Rita: An unwelcome visitor to PPG industries in Lake Charles, Louisiana," *Journal of Hazardous Materials,* vol. 159, pp. 58-60, 2007.

[70] H. K. Kytomaa, P. Boehm, J. Osteraas, B. Haddad, J. Hacker, L. Gilman, E. Jampole, P. Murphy and S. Souri, "An integrated method for

quantifying and managing extreme weather risks and liabilities for industrial infrastructure and operations," *Process Safety Progress,* vol. 38, p. e12087, 2019.

[71]   R. E. Sanders, "Expect the unexpected when thinking extreme weather," *Process Safety Progress,* vol. 38, p. e12082, 2019.

[72]   M. S. Schmidt, "Rare but conceivable: Determining the likelihood of meteors and other infrequent events," *Process Safety Progress,* vol. 38, p. e12090, 2019.

[73]   CSB, "Explosion and Fire, First Chemical Corporation," US Chemical Safety and Hazard Investigation Board, Report No. 2003-01-1-MS, Washington, DC USA, 2003.

[74]   D. F. Schneider, "Plant Power Failures: Case Study of Indirect Effects on a Worldscale Olefins Plant," Stratus Engineering, Inc., League City, Texas, USA, 1998.

[75]   R. S. Ettouney and M. A. El-Rifai, "Explosion of ammonium nitrate solutions, two case studies," *Process Safety and Environmental Protection,* vol. 90, pp. 1-7, 2012.

[76]   CCPS, More Incidents that Define Process Safety, Hoboken, NJ USA: John Wiley & Sons, In Publication, 2019.

[77]   H. Meshkati, "Onagawa: The Japanese Nuclear Plant that Didn't Melt Down on 3/11," The Bulletin of the Atomic Scientists, Chicago, IL USA, 2014.

[78]   CSB, "Organic Peroxide Decomposition, Release, and Fire at Arkema Crosby Following Hurricane Harvey Flooding," US Chemical Hazards and Safety Board, Washington, D.C. USA, 2018.

[79]   CCPS, Guidelines for Safe Automation of Chemical Processes, Second Edition, Hoboken, NJ USA: John Wiley & Sons, 2017.

[80] CCPS Beacon, "What is a Safety Instrumented System?," Center for Chemical Process Safety, Beacon, New York, NY USA, July, 2009.

[81] D. White, "Wrong Pasadena," *Industrial Fire World,* vol. 24, no. 5, pp. 6-20, 2009.

[82] R. A. McConnell, "The Use of Slam Shut Valves on LCA Plants," *Process Safety Progress,* vol. 16, no. 2, pp. 61-68, 1997.

[83] CSB, "Chlorine Release, Report No. 2002-05-I-MO," US Chemical Safety and Hazard Investigation Board, Washington, D.C. USA, 2003.

[84] CSB, "Honeywell International, Inc., Report No. 2003-13-I-LA," U.S. Chemical Safety and Hazard Investigation Board, Washington, DC USA, 2005.

[85] B. K. Vaughen, "An Approach for Teaching Process Safety Risk Engineering and Management Control Concepts Using AIChE's Web-based Concept Warehouse," *Process Safety Progress,* vol. 38, no. 2, pp. 1-18, 2019.

[86] Munich RE, "Machinery and equipment damage during first start-up, testing and commissioning: A guide to loss prevention; HSB-LCE-RGN-13 Rev. 0 Date: 19/02/2016," London, UK, 2016.

[87] S. Mukherjee, "Preparations for Initial Startup of a Process Unit," *Chemical Engineering; www.che.com,* no. January, pp. 36-42, 2006.

[88] CCPS, Guidelines for Defining Process Safety Competency Requirements, Hoboken, NJ USA: John Wiley & Sons, 2015.

[89] I. J. Shin, "Loss Prevention at the startup stage in process safety management: From distributed cognition perspective with an accident case study," *Journal of Loss Prevention in the Process Industries,* vol. 27, pp. 99-113, 2014.

[90] CCPS Beacon, "Air Power!," Center for Chemical Process Safety, Beacon, New York, NY USA, September 2013.

[91] CSB, "Urgent Recommendations (includes ConAgra and Kleen Energy incidents)," US Chemical Safety and Hazard Investigation Board, Washington, D.C. USA, 2010.

[92] CCPS Beacon, "Flammable Vapor Release Hazards in Congested Areas," Center for Chemical Process Safety, New York, NY USA, August 2011.

[93] D. C. Hendershot, "Lessons from Human Error Incidents in Process Plants," *Process Safety and Environmental Protection,* vol. 84, no. B3, pp. 174-178, 2006.

[94] E. Seba, "Insight: In hours, caustic vapors wreaked quiet ruin on biggest U.S. Refinery," Reuters (reuters.com), Port Arthur, Texas, July 24, 2012.

[95] K. Bloch, Rethinking Bhopal: A Definitive Guide to Investigating, Preventing, and Learning from Industrial Disasters, New York, NY USA: Elsevier, 2016.

[96] B. K. Vaughen, "Three decades after Bhopal: What we have learned about effectively managing process safety risks," *Process Safety Progress,* vol. 34, pp. 345-354, 2015.

[97] B. K. Vaughen and K. Bloch, "Use the Bow Tie Diagram to Help Reduce Process Safety Risks," *Chemical Engineering Progress,* vol. 112, no. 12, pp. 30-36, 2016.

[98] A. L. Sepeda, "Understanding Process Safety Management," *Chemical Engineering Progress,* vol. 106, no. 8, pp. 26-33, 2010.

[99] P. R. Amyotte, S. Berger, D. W. Edwards, J. P. Gupta, D. C. Hendershot, F. I. Khan, M. S. Mannan and R. J. Willey, "Why major accidents are still

occuring," *Current Opinion in Chemical Engineering,* vol. 14, pp. 1-8, 2016.

[100] CCPS, Essential Practices for Creating, Strenghtening, and Sustaining Process Safety Culture, Hoboken, NY USA: John Wiley & Sons, 2018.

[101] P. R. Robinson, "Safety and the Environment, Chapter 2," in *Springer Handbook of Petroleum Technology, 2nd Edition,* springer.com, Springer International Publishing AG, 2017, pp. 86-143.

[102] CCPS, Layer of Protection Analysis (LOPA): Simplified Risk Assessment, Hoboken, NJ USA: John Wiley & Sons, 2001.

[103] CCPS, "Process Safety Metrics: Guide for Selecting Leading and Lagging Indicators, Version 3.2," Center for Chemical Process Safety, New York, NY USA, 2019.

[104] CCPS, Conduct of Operations and Operational Discipline, Hoboken, NJ USA: John Wiley & Sons, 2011.

[105] B. K. Vaughen and T. A. Kletz, "Continuing Our Process Safety Management (PSM) Journey," *Process Safety Progress,* vol. 31, no. 4, pp. 337-342, 2012.

[106] S. Z. Halim and M. S. Mannan, "A journey to excellence in process safety management," *Journal of Loss Prevention in the Process Industries,* vol. 55, pp. 71-79, 2018.

[107] CCPS, "Vision 20/20," Center for Chemical Process Safety, 2019. [Online]. Available: www.aiche.org/ccps/resources/vision-2020.

[108] B. K. Vaughen, "Understanding and managing risk during transient operations," in *2020 AIChE/CCPS Spring Meeting,* Houston, 2020.

[109] CCPS, Incidents that Define Process Safety, Hoboken, NJ USA: John Wiley & Sons, 2009.

[110] F. Khan and Editor, Methods in Chemical Process Safety, Volume I, Cambridge, MA USA: Academic Press, Elsevier, 2017.

[111] CSB, "CSB Investigations," US Chemical Safety and Hazard Investigation Board, 2019. [Online]. Available: www.csb.gov/.

[112] CSB, "E.I. DuPont de Nemours & Co., Inc.: Methyl Chloride Release, Oleum Release, Phosgene Release, Report No. 2010-6-I-WV," US Chemical Safety and Hazard Investigation Board, Washington, D.C. USA, 2011.

[113] CSB, "ExxonMobil Torrance Refinery, No. 2015-02-I-CA," U.S. Chemical Safety and Hazard Investigation Board, Washington, D.C. USA, 2017.

[114] CSB, "Fire at Kuraray America EVAL Plant, Factual Investigative Report," U.S. Chemical Safety and Hazard Investigation Board, Washington, DC USA, 2018.

[115] CCPS, "CCPS Process Safety Beacon," Center for Chemical Process Safety, 2019. [Online]. Available: www.aiche.org/ccps.

[116] CCPS Beacon, "Can a water pump explode?," CCPS, New York, NY USA, August, 2013.

[117] CCPS Beacon, "What if That "Wrong" Instrument Reading is Correct?," Center for Chemical Process Safety (CCPS), Beacon, New York, NY USA, April, 2019.

[118] EPA, "EPA Chemical Accident Investigation Report: Tosco Avon Refinery; EPA550-R-98-009," US EPA, Washington, D.C. USA, 1998.

[119] EPSC, "Release from Flare System, EPSC Learning Sheet," European Process Safety Centre (EPSC), www.epsc.be, Frankfurt am Main, Germany, May, 2019.

[120] G. W. Hampton and P. R. Robinson, "Controlling Hydrocracker Excursions, PD-11-01," in *NPRA Q&A and Technology Forum, Plant Automation & Decision Support*, San Antonio, TX USA, 2011.

[121] K. Weick and K. Sutcliffe, Managing the Unexpected: Sustained Performance in a Complex World, 3rd. Edition, Hoboken, NJ USA: Jossey-Bass, a Wiley Brand, 2015.

[122] IEE, "Report on possible Health, Safety and Environmental (HSE) hazards from biomass gasification plants, Deliverable 9," Intelligent Energy - Europe (IEE), Austria, 2007.

[123] CCPS, Guidelines for Integrating Management Systems and Metrics to Improve Process Safety Performance, Hoboken, NJ USA: John Wiley & Sons, 2016.

[124] CCPS, Recognizing and Responding to Normalization of Deviance, Hoboken, N. J.: John Wiley & Sons, 2018.

[125] CCPS, Recognizing Catastrophic Incident Warning Signs in the Process Industries, Hoboken, NJ USA: John Wiley & Sons, 2012.

Index